3000 Questions & Answers for HVAC/R Licensing Examinations

James L. Dundas

NEW ENGLAND INSTITUTE
OF TECHNOLOGY
LEARNING RESOURCES CENTER

Published By
BNP Business News Publishing Company
Troy, Michigan

Copyright © 1989
Business News Publishing Company

All rights reserved. No part of this book may be reproduced or transmitted in any form or by any means—electronic or mechanical, including photocopying, recording or by any information storage and retrieval system—without written permission from the publisher, Business News Publishing Company.

Administrative Editor: Phillip R. Roman
Copy Editor: Jacqueline Smith

Library of Congress Cataloging in Publication Data
Dundas, James Lee.
   3000 questions & answers for HVAC/R licensing examinations.
   I. Heating--Examinations, questions, etc.
2. Ventilation--Examinations, questions, etc.
3. Air conditioning--Examinations, questions, etc.
4. Refrigeration and refrigerating machinery--Examinations, questions, etc.
I. Title. II. Title: Three thousand questions and answers for HVAC/R licensing examinations.
TH7015.D86   1989   697'.0076   89-17474
ISBN 0-912524-51-0

Printed in US
7 6 5 4 3

CONTENTS

SECTION ONE _____ 1
    Refrigeration and Air Conditioning Questions

SECTON TWO _____ 165
    Heating Questions

ANSWERS
    Section One _____ 325
    Section Two _____ 329

Disclaimer

This book is only considered to be a general guide. The author and publisher have neither liability nor can they be responsible to any person or entity for any misunderstanding, misuse or misapplication that would cause loss or damage of any kind, including material or personal injury, or alleged to be caused directly or indirectly by the information contained in this book.

PREFACE

In the field of heating, refrigeration and air conditioning, competency is a very important subject. Municipalities throughout the country test the skills and knowledge of individuals for the purpose of issuing various types of licenses. This testing includes general knowledge in the subject area as well as local and national codes. The following test questions are very similar in nature to those found on tests throughout the country. However, it is necessary that an individual check with the municipality of which a license is desired for information concerning local codes and qualifications.

When taking a test, you may not find what you think is the correct answer. In this instance, it is then important that you select the most correct answer of those provided.

—James Lee Dundas, Professor
Macomb Community College
Warren, Michigan

ABBREVIATIONS

A—amps
AC—alternating current
bip—black iron pipe
Btu—British Thermal Units
cfm—cubic feet per minute
db—dry bulb temperature
DC—direct current
fpm—feet per minute
ft, '—foot or feet
ft^2—square feet
ft^3—cubic feet
ga—gauge
gn—grains
gpm—gallons per minute
ID—inside diameter
in, "—inch
kcal—kilocalories
kW—kilowatts
kWh—kilowatt hours
lbs, #—pounds
LP—liquid propane
mv—millivolts
OD—outside diameter
ppm—parts per million
psi—pounds per square inch
psia—pounds per square inch absolute
psig—pounds per square inch gauge
rh—relative humidity
rpm—revolutions per minute
SSU—Saybolt Seconds Universal
um—universal microns
V—volts
wb—wet bulb temperature
°C—degrees Centigrade
°F—degrees Fahrenheit
" wc—inches water column

More books by Business News Publishing

How To Design Heating/Cooling Comfort Systems
Modern Soldering and Brazing Techniques
The MSAC Service Hot Line Handbooks
How to Make it in The Service Business
Understanding Commercial Ice Makers
Electronic HVAC Controls Simplified
Simplified Cascade System Servicing
Systematic Commercial Refrigeration
Starting in Heating and A/C Service
The A/C Cutter's Ready Reference
HVAC/R Reference Notebook Set
Valve Selection & Service Guide
Refrigeration Licenses Unlimited
The Schematic Wiring Book Set
Service Agreement Dynamics
Schematic Wiring Simplified
Boiler Operator's Dictionary
Industrial Refrigeration
Basic Refrigeration   Boilers Simplified
Humidity Control   Smart Contracting
3000 Questions & Answers for HVAC/R Licensing Exams
How To Solve Your Refrigeration & A/C Service Problems
How To Close-A Guide to Residential Heating/Cooling Sales

For more information on these and other titles, send for a free brochure.

BNP Business News Publishing Company
Book Division, P.O. Box 2600, Troy, MI 48007

SECTION ONE

AIR CONDITIONING & REFRIGERATION

1. According to some refrigeration codes, a self-contained system which has been assembled and tested prior to its installation and which is installed without connecting any refrigerant containing parts is claimed as:
   a) a unit system
   b) a remote system
   c) an indirect system
2. Solenoid valves are generally installed in the:
   a) suction line
   b) water line
   c) liquid line before the TX valve
3. The refrigerant methyl chloride:
   a) can be used for air conditioning
   b) cannot be used for air conditioning
   c) can be used anywhere
4. The maximum setting for the pressure limiting device shall not exceed:
   a) 60% of the setting of the pressure relief valve
   b) 90% of the setting of the pressure relief valve
   c) any setting under 300 psig
5. A part used solely for lowering the temperature of water utilizing the evaporative principle is:
   a) an evaporator
   b) an evaporative condenser
   c) a cooling tower
6. TX valves are normally factory set for:
   a) 5 to 15 degrees of superheat
   b) 22 to 30 degrees of superheat
   c) 0 to 10 degrees of superheat
7. A fusible plug is permitted on the:
   a) low-side of the system only
   b) high-side of the system only
   c) both high- and low-side

8. An increase in pressure on a liquid will cause the boiling point of the liquid to:
   a) decrease
   b) increase
   c) remain the same
9. The minimum height that refrigeration piping can be installed when run across an open space is:
   a) 6 feet
   b) 7½ feet
   c) 8 feet
10. Automatic condenser water regulating valves are operated by:
    a) hand
    b) rise in temperature
    c) rise in head pressure
11. A temprite system using methyl chloride should never use:
    a) an oil separator
    b) a rack pressure valve
    c) aluminum parts
12. A system containing sulphur dioxide:
    a) would never be installed in a hospital
    b) can be installed in any type of occupancy
    c) can be installed in some hospitals with certain reservations
13. Some refrigeration codes do not require a pressure relief valve or a fuse plug on a pressure vessel not exceeding:
    a) an internal gross volume of 3 ft^3
    b) an internal diameter of 3"
    c) 30 psig working pressure
14. With an automatic expansion valve, the pressure in the evaporator with the machine running is:
    a) almost always constant
    b) varies with the load in the cooling compartment
    c) pressure varies with temperature leaving the evaporator
15. Lines that are to be covered must:
    a) be inspected before covering
    b) be soft copper in conduit
    c) have all joints brazed with 1,000°F solder
16. The temperature above which a vapor cannot be condensed regardless of the pressure imposed upon it is:
    a) thermodynamic process
    b) a non-condensable
    c) critical temperature
17. All pressure relief devices ½-inch and over for refrigerant containing pressure vessels shall be marked with the data required in:
    a) section VIII of the ASME boiler and pressure vessel code
    b) Underwriters Laboratory

c) the safety valve code
18. A high side float is most always found on systems:
   a) having one evaporator and one float
   b) of domestic type only
   c) having numerous evaporator and liquid controls
19. If used, a defrost cycle generally occurs when:
   a) the system is in normal operation
   b) the system is stopped or on defrost cycle
   c) the demand for refrigeration is at a maximum
20. The frost line is not affected by:
   a) the charge in the system
   b) the superheat setting of the TX valve
   c) the solenoid valve
21. In a counter-flow condenser, the:
   a) warmest part of one fluid meets the warmest part of the other fluid
   b) warmest part of one fluid meets the coldest part of the other fluid
   c) automatic cooling water valve is installed on the outlet side of the condenser
22. A device used to decrease the starting load of the compressor is:
   a) a capacitor
   b) an equalizer
   c) an unloader
23. A relief valve protecting a pressure vessel from overpressure should start to function at:
   a) a pressure not exceeding the design working pressure of the vessel
   b) the refrigerant field leak test pressure of the system
   c) the pressure corresponding to the cut-off of the high pressure control
24. A substance contains the most heat when it is in the:
   a) gaseous state
   b) liquid state
   c) solid state
25. Some refrigeration codes allow the discharge of the pressure relief device on the compressor to be:
   a) vented to the low-side of the system
   b) vented to the high-side of the system
   c) dumped into the city sewer
26. An increase in discharge pressure would cause an automatic regulating valve:
   a) to open more
   b) to close off
   c) to restrict the flow of water
27. The refrigerant having the highest latent heat is:
   a) R-22
   b) R-12
   c) ammonia

28. An agent often used with methyl chloride to call attention to the refrigerant is:
    a) SO_2
    b) Acrolein
    c) methylene chloride
29. An accumulator, if used, will be located in the:
    a) low-side
    b) high-side
    c) rear bank of tubes
30. A system using an automatic expansion valve:
    a) must be hand operated
    b) can be thermostatically operated
    c) can be operated with a hi-low switch
31. Some refrigeration codes require lines that are to be covered with concrete or earth. These lines must be:
    a) made of soft copper
    b) encased in rigid conduit
    c) covered with earth or concrete after inspection
32. Oil will absorb refrigerant more readily when it is:
    a) hot
    b) warm
    c) cold
33. Discharge lines from air conditioners shall:
    a) be directly connected to the waste or sewer system
    b) terminate over and above a trapped and vented plumbing fixture
    c) be connected to any convenient opening
34. The fusible plug shall be connected directly to the:
    a) evaporator
    b) pressure vessel
    c) anywhere on the system
35. The use of compressor high-pressure gas in the evaporator for the removal of frost is known as:
    a) hot gas defrost
    b) leak testing
    c) direct defrost
36. An instrument to measure specific gravity is a:
    a) psychrometer
    b) gravitational device
    c) hydrometer
37. A pressure vessel shall bear ASME stamping when it exceeds:
    a) 6" internal diameter
    b) 12" internal diameter
    c) 18" internal diameter

38. Increasing the superheat setting of a TX valve will:
    a) increase the refrigerating capacity of the unit
    b) not change the refrigerating capacity of the unit
    c) decrease the refrigerating capacity of the unit
39. The by-pass safety valve on a refrigerating system is located between:
    a) check valves in the discharge line and the condenser
    b) suction and discharge valves on the compressor
    c) the evaporator and the compressor
40. Latent heat is defined as the:
    a) heat added by the compressor to the suction gas
    b) heat which is associated with a change in temperature that can be measured with a thermometer
    c) amount of heat required to bring about melting or vaporizing
41. 50/50 solder joints are:
    a) permitted on systems containing less than 6 lbs of Group 1 refrigerants
    b) not permitted on any system
    c) permitted on NH_3 systems
42. The purpose of a pressure limiting device is to:
    a) maintain a constant suction pressure to the compressor
    b) stop the compressor in case of over pressure
    c) maintain constant cooling water pressure to the condenser
43. Although a cooling tower is a water-conserving device, some make-up water is necessary to compensate for the loss. This percentage of make-up water is:
    a) 1 to 3 %
    b) 6 to 10 %
    c) 15 to 20 %
44. An instrument used to measure relative humidity is a:
    a) humidifier
    b) hydrometer
    c) psychrometer
45. Dirty condenser tubes could lead to:
    a) dirt accumulation on evaporator tubes
    b) decrease in head pressure
    c) increase in head pressure
46. Excessive head pressure in a refrigeration system is usually caused by:
    a) condenser water failure
    b) insufficient refrigerant
    c) a head gasket that is too thick
47. A gas tight joint obtained by joining of metal parts with metallic mixtures or alloys which melt at temperatures above 1000°F and less than the melting temperatures of the joined parts is a:
    a) leaded joint
    b) soldered joint
    c) brazed joint

48. How many Btu must be added to change a ton of ice at 32° into water at 32°?
    a) 500,000
    b) 176,000
    c) 288,000
49. When installing a relief valve on a pressure vessel, it should be placed:
    a) above the liquid level
    b) below the liquid level
    c) anywhere on the vessel
50. The pressure-type condenser water valve is actuated by:
    a) high-side pressure
    b) low-side pressure
    c) intermediate pressure
51. Refrigerating effect is:
    a) weight of refrigerant circulated per minute
    b) amount of heat absorbed by the refrigerant
    c) the horsepower per ton of refrigeration
52. When required, a check valve would be located:
    a) in the suction line to the compressor
    b) between the high and low sides of the compressor
    c) in the discharge line of the compressor
53. The maximum size pressure vessel which may use a fuse plug as the sole means of relieving excessive pressure is a gross volume:
    a) of 3 ft^3 or less
    b) of 3 ft^3 or over, but less than 10 ft^3
    c) over 10 but less than 15 ft^3
54. A control device for regulating the flow of liquid refrigerant into an evaporator actuated by changes in pressures in the evaporator is:
    a) a thermostatic expansion valve
    b) an automatic expansion valve
    c) a solenoid valve
55. When concealing copper tubing, you provide removable access plates on:
    a) all concealed joints
    b) only soft solder joints
    c) only flaired joints
56. Heat is transferred in which of the following three manners:
    a) radiation, conduction, convection
    b) radiation, conduction, absorption
    c) convection, conduction, refraction
57. When installing refrigeration equipment on roofs, the responsibility for providing access to the equipment rests with the:
    a) contractor
    b) owner of the equipment
    c) owner of the building

58. Pressure limiting devices shall be provided on:
   a) water-cooled systems only
   b) all systems containing more than 20 lbs of refrigerant and on all water-cooled systems
   c) any systems over 2 horsepower
59. A device used to rid a system of non-condensable gases is called a:
   a) purger
   b) classifier
   c) drier
60. When installing a water regulator valve on a refrigerating system, locate the valve:
   a) on the water inlet
   b) on the water outlet
   c) anywhere on the system
61. When replacing a vessel on a refrigerating system, it is important that the vessel:
   a) be properly identified regarding necessary approval
   b) have the same number of openings as the original vessel
   c) have the gauge metal or thicker
62. The reason for a frosted compressor using a TXV could be the result of:
   a) the expansion valve passing too much refrigerant
   b) a restricted liquid line
   c) high-pressure control not cutting off
63. What size pressure vessel requires the installation of a pressure relief device?
   a) 3 to 10 ft^3
   b) over 6" diameter
   c) over 50 lbs capacity
64. Latent heat involves a change between the liquid and solid states of a substance:
   a) latent heat of fusion
   b) latent heat of evaporation
   c) latent heat of sublimation
65. It is considered safe practice to install a rupture disc:
   a) with no markings on the disc
   b) on the upstream (inlet) side of the relief valve
   c) on the downstream (outlet) side of the relief valve
66. A device installed in a compressor head and which is used to protect the compressor from damage due to liquid slugging is a:
   a) safety head
   b) jacketed cylinder
   c) cylinder head
67. The type of expansion valve that could not be used on a flooded evaporator is:
   a) a low-side float
   b) an automatic expansion valve
   c) a thermostatic expansion valve

68. A "liquid seal" in a refrigeration system is considered necessary to:
    a) stop refrigerant leaks around the shaft
    b) insure a flooded evaporator
    c) prevent hot gas from entering the evaporator
69. A heat exchanger, used with R-12 systems, is for the purpose of:
    a) decreasing suction superheat
    b) increasing suction superheat
    c) reducing flash gas
70. A controlling device for regulating the flow of refrigerant into a cooling unit, actuated by changes in superheat of the vapor leaving the unit, is a:
    a) thermostatic expansion valve
    b) hand expansion valve
    c) superheat control
71. Some refrigeration codes require that the discharge of a fusible plug be on the outside of the building on all systems:
    a) used in institutional occupancies
    b) containing more than 6 lbs of Group II or Group III refrigerants
    c) installed in your city
72. The transmission of heat through intervening substances without heating these substances is:
    a) conduction
    b) convection
    c) radiation
73. When two or more condensers are served in parallel from a common water supply with a shut-off valve on the outlet side of the condenser, a check valve shall be installed:
    a) upstream from all supply connections
    b) downstream from all supply connections
    c) any place on the water line
74. A device having a pre-determined melting temperature member for the relief of pressure is a:
    a) relief valve
    b) fusible plug
    c) fusible link
75. If Freon comes in contact with an open flame, it could become dangerous because it could break down and form:
    a) hydrogen gas
    b) sulphur dioxide
    c) phosgene
76. An atmospheric condenser is similar to:
    a) a shell and tube condenser
    b) a shell and coil condenser
    c) an evaporative condenser without a blower

77. A thermostatic expansion valve will control the evaporator:
   a) pressure
   b) temperature
   c) quantity of refrigerant
78. Joints on piping located in an air duct of an air-conditioning system shall be constructed to stand a temperature of:
   a) 450°F to 1000°F
   b) 1,000°F or over
   c) 15,000°F or over
79. The quantity of heat necessary to raise the temperature of 1 lb of water 1°F is known as:
   a) thermometer
   b) Btu
   c) thermocouple
80. Soft annealed copper tubing used with Group II refrigerants shall:
   a) be inspected after installation
   b) have rigid or flexible metal enclosures
   c) be 95/5 solder joints in institutional occupancies
81. An equalizer connection to a TX valve on a large evaporator coil will:
   a) tend to starve the coil under a heavy load
   b) tend to flood the coil under a heavy load
   c) permit control with 1 or 2 degrees of superheat
82. The defrost system that uses an anti-freeze solution in the heat exchanges is a:
   a) thermobank
   b) water spray
   c) hot gas by-pass
83. The growth of molds or bacteria on foods can be stopped by:
   a) lowering the temperature
   b) lowering the temperature below the freezing point of the product
   c) storing under 32°F
84. The type of condenser that gives the best separation of liquid and vapor is the:
   a) double pipe
   b) evaporative
   c) shell and tube
85. A fractional horsepower motor that gives the best starting torque is:
   a) induction with one set of windings
   b) capacitor split-phase motor
   c) resistance split-phase
86. At least 2 gas masks shall be provided when the quantity of refrigerant exceeds:
   a) 100 lbs of Group II refrigerant
   b) 100 lbs of Group III refrigerant
   c) 300 lbs of refrigerant

87. The rate of heat interchange (12,000 Btu/hr or 200 Btu/min) is:
    a) latent heat
    b) a ton of refrigeration
    c) refrigerating effect
88. A system having the evaporator in contact with the material or space refrigerated is called a:
    a) flooded system
    b) an indirect system
    c) direct system
89. A higher heat transfer rate is usually attained with a:
    a) direct system
    b) dry expansion system
    c) flooded system
90. The ripening process of food:
    a) releases heat which is known as "vital heat"
    b) is caused by bacteria
    c) is slowed down by warm temperature
91. A collecting or storage chamber for low-side liquid refrigerant is called:
    a) an absorber
    b) an accumulator
    c) a regenerator
92. The function of the compressor is to:
    a) compress the liquid refrigerant so it will vaporize
    b) circulate the refrigerant
    c) compress the refrigerant vapor above the condensing point and to help circulate refrigerant
93. Which of the following occupancies has the most stringent code requirements?
    a) public assembly
    b) commercial
    c) institutional
94. Heat always travels:
    a) from a warmer to a colder substance
    b) from a colder to a warmer substance
    c) upstream
95. What can be used for building up a pressure in a refrigeration system for testing purposes?
    a) air
    b) oxygen
    c) CO_2
96. A check valve is required in the discharge line from a compressor:
    a) when more than 40 lbs of Group II or Group III refrigerant is used
    b) when more than 100 lbs of refrigerant is used
    c) on all systems

97. A compressor relief valve normally discharges to:
    a) the low-side of the system
    b) ambient air
    c) the outside air
98. Dry bulb temperature is a measure of the:
    a) dew point of air
    b) total heat of air
    c) sensible heat of air
99. Cross-heads are used in:
    a) vertical single-acting machines
    b) both vertical single-acting and horizontal double-acting compressors
    c) horizontal double-acting machines
100. The flooded evaporator of a flooded system can be correctly defined as cooling coils that:
    a) are normally full of liquid refrigerant
    b) are normally immersed in some secondary cooling medium such as brine
    c) continuously flood back to the compressor, causing liquid slugging
101. Of the following items, the one that would have the greatest effect upon the performance of an evaporative condenser would be:
    a) ambient temperature
    b) dry bulb (db) temperature
    c) wet bulb (wb) temperature
102. In the operation of a thermostatic expansion valve, which is true:
    a) the pressure of the bulb fluid acts to open the valve
    b) the pressure of the refrigerant in the evaporator acts to open the valve
    c) the spring acts with the evaporator pressure to keep the valve open in relation to the amount of superheat setting
103. An accepted method of preventing frost formation on air blast freezers is by the use of:
    a) hot gas
    b) electric heaters
    c) brine spray
104. Compression ratio is:
    a) initial pressure divided by final pressure
    b) initial volume divided by final volume
    c) final volume divided by initial volume
105. Of the three given compressors, which would not meet accepted standards?
    a) 6" x 6" at 720 rpm
    b) 3" x 3" at 1500 rpm
    c) 6" x 6" at 1500 rpm
106. The liquid refrigerant that is chilled to evaporator temperature in the expansion coils:
    a) expands after its temperature is lowered
    b) contracts after its temperature is lowered

c) doesn't change in volume after its temperature is lowered
107. Heat exchangers, as used with a refrigeration system, are for the purpose of:
    a) decreasing suction superheat
    b) increasing suction superheat
    c) reducing flash gas
108. Methyl Chloride is:
    a) toxic only
    b) flammable only
    c) both toxic and flammable
109. The high pressure cut-out must be connected to the compressor with:
    a) an intervening stop valve
    b) no intervening stop valve
    c) an intervening stop valve of the "T" or lever handle type
110. In a centrifugal compressor, the high velocity of the gas as it leaves the impeller is transformed into static pressure by the:
    a) damper
    b) pressurestat
    c) stationary diffuser
111. The most suitable refrigerant to use in applications of 110°F below zero is:
    a) R-12
    b) ammonia
    c) carbon dioxide
112. The capacity of an ammonia compressor operating at 24 psi suction, as compared to the same compressor operating at 4 psi, is:
    a) about the same
    b) twice as much
    c) six times as much
113. The rate at which work is done is defined as:
    a) energy
    b) motion
    c) power
114. Eliminators, as used on an evaporative condenser, are for the purpose of:
    a) directing the flow of air over tubes
    b) directing the flow of water over tubes
    c) preventing carryover of water with discharge air
115. A low-side float valve will maintain a:
    a) liquid level in an evaporator which will correspond to load and pressure
    b) constant liquid level in the evaporator regardless of the load and pressure
    c) constant pressure in the evaporator
116. The water sides of the tubes of a vertical shell and tube condenser can be cleaned mechanically:
    a) only when the unit is out of service
    b) when the unit is in service
    c) on weekends only

117. The ability of oil to block the passage of electrical current is referred to as its:
    a) KV rating
    b) power factor rating
    c) ampere rating
118. Lower head pressures and discharge temperatures can be obtained by use of:
    a) compound compression
    b) single stage compression
    c) evaporative condensers
119. Silica gel activated alumina are classed as:
    a) saturators
    b) absorbents
    c) adsorbents
120. An external equalizer, if connected to a thermostatic expansion valve:
    a) equalizes the pressure on the inlet and outlet sides of the valve
    b) compensates for excessive pressure drops in the cooling coils
    c) equalizes the temperature of the refrigerant with that of the air surrounding the coil
121. Vertical shell and tube condensers are generally used to condense:
    a) ammonia
    b) carbon dioxide
    c) freon
122. The term "coefficient of performance", as applied to refrigeration, means:
    a) performance of the system due to expansion
    b) the heat input as compared to the heat transfer
    c) the ratio of refrigeration produced to the work supplied
123. The expanding fluid most adaptable for use in refrigeration thermometers for temperatures above -38°F is:
    a) alcohol
    b) red oil
    c) mercury
124. A dehumidifier is a device used to:
    a) decrease humidity
    b) increase humidity
    c) increase the moisture content of the air
125. A high-pressure cut-out shall connect directly to the:
    a) condenser
    b) discharge stop valve
    c) compressor
126. In a brewery, carbon dioxide is:
    a) used in the refrigeration system
    b) added to the beer to reduce its temperature
    c) added to the beer to improve its taste

127. Ingredients used in the beer brewing process which must be kept in cold storage are:
    a) hops and yeast
    b) barley and yeast
    c) barley and rice
128. Oil return problems are less common in a system using:
    a) R-22
    b) R-12
    c) R-114
129. An excessive charge of refrigerant in a system utilizing a high-side float valve could result in:
    a) more efficient cooling
    b) the stoppage of refrigerant at the expansion valve
    c) damage to the compressor
130. Other conditions being equal, more condensing water is required when the source of water supply is:
    a) the city water mains
    b) a natural draft cooling tower
    c) an underground well
131. If heat is added to a gas at constant volume, the pressure of the gas will:
    a) increase
    b) decrease
    c) double
132. Wet compression on a booster compressor would be indicated by:
    a) frost on the suction line
    b) a cold discharge line
    c) a warm discharge line
133. The function of an oil regenerator is to:
    a) return the oil carryover back to the generator
    b) purify the oil
    c) separate the refrigerant from the oil
134. Condensing water is somewhat hard and has scale-forming tendencies. The most practical type of condenser would then be a:
    a) double pipe type
    b) vertical shell and tube type
    c) horizontal shell and tube type
135. A refrigeration pressure relief device may discharge into the low pressure side providing the:
    a) low-side capacity is equal to or greater than high-side
    b) low-side is equipped with fusible plugs
    c) low-side is equipped with a pressure relief device vented to atmosphere
136. Mechanical horsepower represents the completion of 33,000 ft/lb of work in:
    a) one hour
    b) one minute

c) one day
137. The lineal clearance of a compressor piston should be:
    a) greater at the head end of the piston
    b) greater at the crank end of the piston
    c) equal for both sides of the piston
138. Change of state of a substance directly from a solid to a gas without appearance of liquid is known as:
    a) condensation
    b) sublimation
    c) gasification
139. Steam engines are primarily designed to drive a:
    a) horizontal double-acting compressor
    b) vertical single-acting compressor
    c) centrifugal compressor
140. For all practical purposes, the oil separator for an ammonia system should be placed near the:
    a) outlet from the compressor
    b) inlet to the condenser
    c) outlet from the condenser
141. Gaskets on flanged joints on an ammonia system, high pressure side, must be:
    a) rubber
    b) lead
    c) asbestos composition
142. Leaks past valves and piston rings within the compressor cylinder can be pinpointed by:
    a) using a sounding tube
    b) taking an indicator diagram
    c) overhauling the compressor
143. The temperature at which oil no longer flows is called the:
    a) viscosity
    b) floc point
    c) pour point
144. Proper operating procedure requires a compressor to remove the vapor in the low-side of the system:
    a) as fast as the evaporator produces it
    b) faster than the evaporator produces it
    c) much slower than the evaporator produces it
145. Sweat or solder copper alloy fittings are used with:
    a) soft copper tubing
    b) hard copper tubing
    c) steel tubing
146. For a system of given size, a large solenoid valve would be required for the:
    a) control line
    b) liquid line

c) suction line
147. The agent best suited to extinguish a small electrical fire in a machinery room would be:
   a) a foam solution
   b) carbon dioxide
   c) water
148. A capillary tube will best control the refrigerant flow when used with a:
   a) back-pressure control switch
   b) temperature control switch
   c) high-pressure cut-out switch
149. The temperature at which sublimation of solid carbon dioxide occurs at atmospheric pressure is approximately:
   a) 0°F
   b) -50°F
   c) -109°F
150. The specific heat of a substance is higher when that substance is:
   a) above freezing temperature
   b) below freezing temperature
   c) anything but water
151. Compressor capacity control is best achieved by using:
   a) a compound compressor
   b) a booster compressor
   c) valve unloaders
152. To maintain a proper heat balance in a brewery refrigeration plant:
   a) only electric prime movers should be considered
   b) both steam and electric prime movers should be considered
   c) only steam prime movers should be considered
153. The lowest freezing point that can be obtained with a sodium chloride brine is approximately:
   a) 32°F
   b) 15°F
   c) -6°F
154. There is less danger of flooding the compressor at the start of the cycle when the solenoid valve is placed in the:
   a) liquid line to the evaporator
   b) suction line from the evaporator
   c) expansion valve equalizing line
155. The liquid inlet and outlet lines should be connected to the receiver:
   a) with a check valve in each line
   b) as far apart as practical
   c) as close together as is practical
156. The speed at which a reciprocating compressor may operate is limited by the:
   a) crankshaft rotation speed

b) piston speed
   c) belt speed
157. The eliminator used in conjunction with systems having centrifugal compressors is located in the:
   a) condenser
   b) evaporator (cooler)
   c) compressor
158. An automatically controlled compressor operating intermittently usually requires a drive having:
   a) a high starting torque
   b) a low starting torque
   c) no starting torque
159. Ice formation on the brine side of the cooling coils in an ice-making plant:
   a) decreases the time required for ice manufacturing
   b) increases the time required for ice manufacturing
   c) has no effect on the time required for ice manufacturing
160. The float and mechanism of a high-side float are located in the:
   a) surge drum or accumulator
   b) low-pressure side of the system
   c) high-side of the system
161. The water enters a vertical shell and tube condenser at the:
   a) center
   b) top
   c) bottom
162. Recirculation of the brine in an ice-making field:
   a) increases the time required for making ice
   b) decreases the time required for making ice
   c) considerably decreases the strength of the brine solution
163. If the pressure in a system is less than atmospheric, it's normally measured in:
   a) inches of mercury
   b) inches of water
   c) inches of vacuum
164. The recommended maximum piston speed of a reciprocating compressor is:
   a) 600 fpm
   b) 750 fpm
   c) 1000 fpm
165. Stoppage of flow at the expansion valve caused by moisture in the system is more serious with:
   a) ammonia
   b) freon
   c) dry ice
166. A lower condenser pressure will require:
   a) greater work for the compressor

b) less work for the compressor
c) greater power to drive the compressor

167. The lower the suction pressure on a refrigerant, the:
    a) less horsepower required to handle a given quantity of gas
    b) less volume of gas required per ton of refrigeration
    c) greater the volume of gas required per ton of refrigeration

168. A larger clearance on a compressor will:
    a) decrease volumetric efficiency
    b) increase volumetric efficiency
    c) result in oil slugging

169. In a refrigeration cycle, the greatest pressure drop occurs as the refrigerant passes through the:
    a) evaporator
    b) reducing valve
    c) expansion valve

170. Motors, if used to drive compressors, should be:
    a) two phase
    b) single phase
    c) three phase

171. What size replacement capillary tube is usually recommended?
    a) the next largest size
    b) the next smallest size
    c) the same size

172. When at its critical temperature, a refrigerant has:
    a) a higher liquid than gas density
    b) a higher gas than liquid density
    c) equal densities of gas and liquid

173. The quality of ice in an ice-making plant:
    a) is not affected by the rate of cooling
    b) is best maintained at or near a specified brine temperature
    c) is increased for every Fahrenheit degree of decrease in brine temperature

174. The ratio of the refrigerant gas pumped to the piston displacement of the compressor is called:
    a) volumetric efficiency
    b) a ton of refrigeration
    c) clearance

175. Explosions caused by anaesthetics in hospital operating rooms can be reduced by carrying a:
    a) high relative humidity
    b) low relative humidity
    c) high static spark condition

176. A substance having a boiling point above 32°F at atmospheric pressure:
    a) could not be used as a refrigerant
    b) could be used as a refrigerant

c) could be used but won't produce a temperature below 32°
177. The evaporating temperature or condensation temperature of a fluid is the:
    a) critical temperature
    b) saturation temperature
    c) mean temperature
178. An economizer, as used in a refrigeration system, is identified with:
    a) centrifugal compressors
    b) booster system
    c) flash intercoolers
179. Spherical heads on compressors:
    a) increase volumetric efficiency
    b) permit greater valve area
    c) increase pressure developed
180. The function of a heat exchanger connected to the suction and liquid lines is to decrease the:
    a) liquid temperature
    b) suction temperature
    c) discharge temperature
181. A quicker and higher vacuum can be pumped when using a:
    a) centrifugal compressor
    b) reciprocating compressor
    c) rotary compressor
182. A high suction pressure and low head pressure indicates that the compressor is deficient in:
    a) refrigeration gas
    b) oil charge
    c) pumping capacity
183. A compressor relief valve may discharge to:
    a) the low-side of the system
    b) refrigerant well
    c) condenser inlet
184. Of the following substances, which could be considered the most nearly perfect gas?
    a) H_2O
    b) ammonia
    c) nitrogen
185. Leakage at the shaft seal of a centrifugal compressor is generally:
    a) of little importance to the efficient operation
    b) inward to the compressor
    c) outward to the atmosphere
186. The thickness or thinness of an oil and its resistance to flow at a certain temperature is called:
    a) viscosity

b) floc point
c) pour point
187. If the evaporating coils are located to cool some medium which is circulated in the region to be cooled, the system is called:
   a) an indirect system
   b) a flooded system
   c) a direct system
188. Increasing the velocity of the water flow in a condenser:
   a) increases the capacity of the condenser
   b) decreases the capacity of the condenser
   c) does not affect the capacity of the condenser
189. Gas binding difficulties in condenser operation are experienced to a greater degree with the:
   a) evaporative type
   b) vertical shell and tube type
   c) atmospheric type
190. Liquid Carrene 1, released into the atmosphere, will ordinarily:
   a) change to a solid
   b) remain a liquid
   c) change to a gas
191. Oil separators on ammonia systems are:
   a) always placed as close to the condenser as possible
   b) in some systems next to the compressor because of the higher temperature
   c) never next to the compressor because of the higher temperature
192. A larger condenser capacity for a given floor area may be obtained with the:
   a) vertical shell and tube type
   b) horizontal shell and tube type
   c) vertical shell and coil type
193. The freezing point for R-718 is:
   a) -108°F
   b) 32°F
   c) -40°F
194. When the load on a residential air conditioner decreases, the suction pressure:
   a) goes down and the discharge pressure goes up
   b) goes down and the discharge pressure goes down
   c) goes up and the discharge pressure goes down
195. Water will absorb more ammonia when the water temperature is at:
   a) 50°F
   b) 100°F
   c) 200°F
196. What does an anemometer measure?
   a) temperature
   b) velocity
   c) volts

197. Air will absorb more moisture when it is:
    a) cold and dry
    b) hot and saturated
    c) cold and saturated
198. A pump-out compressor is used to:
    a) evacuate or transfer refrigerant in one or more parts of a system
    b) maintain more than one suction pressure and temperature
    c) separate the refrigerant from the oil
199. To lower the superheat setting of a TXV, it is necessary to:
    a) tighten the spring
    b) loosen the spring
    c) add refrigerant to the bulb
200. The special steel cylinders or drums used for storage/transfer of refrigerants:
    a) should be filled completely full
    b) should not be over-filled
    c) may be filled completely full if provided with a fusible plug
201. What safety device is required on all systems containing more than 20 lbs of refrigerant?
    a) fusible plug
    b) pressure limiting device
    c) spring loaded relief device
202. A pressure relief valve is required on:
    a) compressors over 20 hp capacity
    b) all compressors using Group II or Group III refrigerant
    c) all positive displacement compressors exceeding 50 cubic feet of displacement per minute
203. To properly test the accuracy of a refrigeration thermometer, it should be immersed in a:
    a) thermometer well
    b) steam bath
    c) bath of pure water and ice
204. The freezing point for R-717 is:
    a) -40°F
    b)   32°F
    c) -108°F
205. To change a refrigerant liquid into a gas at its critical temperature and pressure:
    a) latent heat is not absorbed
    b) latent heat must be absorbed
    c) is impossible
206. Increasing the clearance volume of a compressor would result in:
    a) no change in the volumetric efficiency
    b) decrease in the volumetric efficiency
    c) increase in the volumetric efficiency

207. One horsepower of mechanical energy is equal to:
    a) 746 watts
    b) 3,413 watts
    c) 2,513 watts
208. The suction valves in a vertical single-acting ammonia compressor are usually located in the:
    a) piston
    b) safety head
    c) cylinder wall
209. When the temperature of the cooling coil is below the dew point and the temperature of the room is below 32°F:
    a) moisture will condense and freeze on the coil
    b) the coils will sweat
    c) the efficiency of the cooling coil will be increased   considerably
210. The evaporation of the cooling water flowing over an atmospheric condenser:
    a) has no effect on the quantity of water handled
    b) increases the amount of water handled
    c) decreases the amount of water handled
211. Gas masks or helmets approved for use shall bear a marking by the:
    a) Dept. of Buildings and Safety Engineering
    b) Bureau of Mines (U.S. Dept. of the Interior)
    c) Interstate Commerce Commission
212. To convert Fahrenheit degrees into Centigrade degrees:
    a) °C = 5/9 x (°F - 32)
    b) °C = 9/5 x (°F - 32)
    c) °C = 5/9 x (°F + 32)
213. The striking clearance in a vertical single-acting ammonia compressor is approximately:
    a) 1/64 inch
    b) 1/4 inch
    c) 1/2 inch
214. Vapor which is at a temperature higher than the saturation temperature at the existing pressure is called:
    a) superheated vapor
    b) saturated vapor
    c) steam vapor
215. An intercooler used between the cylinders of compound compressors is similar in appearance to a:
    a) shell and tube or surface condenser
    b) double pipe condenser
    c) evaporative condenser
216. A function of the accumulator is to:
    a) prevent the compressor from becoming vapor-bound

b) keep the expansion coils filled with liquid
c) separate the oil from the refrigerant

217. A superheated vapor is a vapor:
    a) free of liquid
    b) saturated with liquid
    c) at a temperature higher than that corresponding to the pressure

218. The main function of a compressor's clearance pocket is to provide:
    a) for piston lineal clearance
    b) for refrigeration capacity control
    c) a recess for the piston rod jam nut

219. The bulb of a thermostatic expansion valve generally contains:
    a) refrigerant
    b) water
    c) mercury

220. An evaporative condenser will operate more efficiently:
    a) on a low-humidity day
    b) on a high-humidity day
    c) if the blower is turned off

221. In making ice, the primary purpose for blowing air into the cans is to:
    a) shorten the time required for freezing
    b) aid in the removal of impurities
    c) make the ice much lighter

222. The refrigeration operator's responsibility to maintain the equipment under his supervision in a clean and accessible manner:
    a) is mandatory in the refrigeration code
    b) is not mandatory in the refrigeration code
    c) should be delegated to the janitor

223. Pumps, when used to agitate the brine in an ice field, are the:
    a) centrifugal type
    b) rotary type
    c) reciprocating type

224. In ammonia plants, the condenser tubes should be composed of:
    a) steel
    b) copper
    c) brass

225. An intercooler is used in conjunction with:
    a) two coolers
    b) two-stage or compound compressors
    c) single-stage compressors

226. Moisture eliminators are generally found on:
    a) evaporative condensers
    b) shell and tube condensers
    c) double-tube or double-pipe condensers

227. Subcooling the liquid between the condenser and the evaporator:
    a) has no effect on the flash gas at the expansion valve
    b) increases the flash gas at the expansion valve
    c) decreases the flash gas at the expansion valve
228. The condenser which performs with the least water consumption is:
    a) the evaporative type
    b) shell and tube type
    c) double pipe type
229. The canisters or cartridges for gas masks that have been used should be renewed:
    a) immediately
    b) within two years
    c) if the seal cannot be replaced
230. As the pressure imposed on a refrigerant increases, the latent heat of evaporization:
    a) remains constant
    b) decreases
    c) increases
231. Assuming that both types of compressors are serving the same system, the stuffing box that will have to withstand the greater pressure would be on the:
    a) vertical single-acting machine
    b) horizontal double-acting machine
    c) centrifugal machine
232. The heat transfer rate per ft^2 of evaporator surface is usually greater with:
    a) a dry expansion system
    b) a flooded system
    c) both saturated and superheated vapor in the system
233. A stop valve, if located between a pressure relief device and a vessel, is:
    a) not permissible under any circumstances
    b) permissible
    c) permissible, providing two or more valves are connected to the stop valve in parallel and only one can be rendered inoperative for testing and repair
234. Freon 11 has the same characteristics as:
    a) Carrene 7
    b) Carrene 1
    c) Carrene 2
235. The statement that heat always flows from the substance of a higher temperature to a substance of lower temperature of its heat content is the definition of:
    a) the second law of thermodynamics
    b) the first law of thermodynamics
    c) the third law of thermodynamics

236. Piston rings with the inside bore eccentric (thus making them thicker at one part than another) have the cut for the joint at the:
    a) center
    b) thinnest part
    c) thickest part
237. It is considered good practice to connect the automatic water regulating valve to the:
    a) city water make-up line
    b) condenser outlet
    c) condenser inlet
238. In case of condenser water failure, which of the following safety devices should operate first?
    a) high-pressure cut-out
    b) low water cut-out switch
    c) pressure relief valve
239. The main advantage of an indirect system has over a direct system is:
    a) increased operating efficiency
    b) overall safety, for the refrigerant can be completely removed from the area being cooled
    c) the requirement for fewer parts
240. To obtain a desired temperature with the greatest efficiency, the back pressure of the system should be maintained:
    a) as high as possible
    b) as low as possible
    c) below atmospheric
241. Indicator diagrams can be taken to check the performance of the:
    a) compressor
    b) expansion valves
    c) automatic purge device
242. Which of the following refrigerants is the most toxic?
    a) the Freons
    b) carbon dioxide
    c) ammonia
243. Other conditions being the same, ice of the highest quality would be made at a temperature of:
    a) 32°F
    b) 0°F
    c) 16°F
244. Other conditions remaining the same, cold air:
    a) can absorb less moisture than hot air
    b) can absorb more moisture than hot air
    c) cannot contain any moisture

245. The refrigerant vapor in a vertical shell and tube condenser:
    a) flows outside the tubes in the shell
    b) flows inside the tubes
    c) neither a or b, as this type condenser is no longer used
246. When determining relative humidity with a psychrometer, the humidity can be obtained:
    a) directly
    b) through the use of a psychrometric chart
    c) in grains per gallon
247. The high-pressure cut-out shall stop the action of the compressor at a pressure not exceeding:
    a) 75% of the design working pressure of the system
    b) 90% of the relief device setting
    c) 90% of the pressure relief device setting, 90% of the refrigerant leak test pressure applied or 90% of the design working pressure of the high-side, whichever is smallest
248. A suitable refrigerant should have a critical temperature and pressure:
    a) well below the condensing temperature and pressure of the system
    b) well above the condensing temperature and pressure of the system
    c) above pressure and below temperature only
249. Evaporative condensers are usually located on building roofs to:
    a) maintain a static head on liquid lines
    b) take advantage of large areas of empty space
    c) take advantage of the cooling effect obtained by the evaporation of the cooling water
250. A device used to test the density or strength of brine is called a:
    a) calorimeter
    b) hydrometer
    c) salinometer
251. Of the refrigerants listed below, the one which makes the most effective fire extinguisher is:
    a) carbon dioxide
    b) sulphur dioxide
    c) Freon 12
252. Other conditions being equal, as the latent heat increases, the:
    a) quantity of refrigerant that must be circulated increases
    b) quantity of refrigerant that must be circulated decreases
    c) sensible heat decreases
253. The type of compressor most suitable for handling large quantities of refrigerant vapor is the:
    a) reciprocating type
    b) centrifugal type
    c) hermetic type

254. When heat is added to a gas at a constant pressure, the gas will:
    a) contract
    b) expand
    c) remain constant
255. Feather valves used in compressors are:
    a) bevel seated
    b) flat seated
    c) spring seated
256. An external equalizer, when used in conjunction with a thermostatic expansion valve, is connected to the:
    a) feeler bulb capillary tube
    b) same side of the power element to which the feeler bulb is attached
    c) side of the power element opposite from that having the feeler
257. The water loss through use of a cooling tower or spray pond runs:
    a) 0%
    b) between 25% and 50%
    c) between 5% and 10%
258. Condenser cooling water having a pH value of 7 is considered:
    a) alkaline
    b) acid
    c) neutral
259. The purpose of an oil safety switch used on a refrigeration system is to stop the compressor in case of:
    a) excessive head pressure
    b) insufficient oil pressure
    c) excessive oil pressure
260. The latent heat of carbon dioxide, when sublimated at atmospheric pressure, is approximately:
    a) 246 Btu/lb
    b) 296 Btu/lb
    c) 346 Btu/lb
261. The standard operating conditions for determining the refrigerating capacity of a machine are:
    a) 86°F condensing & 5°F evaporating temperatures of the refrigerant
    b) 212°F condensing & 32°F evaporating temperatures of the refrigerant
    c) 86°F condensing & 5°F evaporating temperatures of the brine
262. Under certain conditions, it is permissible to charge a unit at the:
    a) condenser
    b) receiver
    c) liquid line
263. 778 ft/lb equals:
    a) one horsepower
    b) one Btu
    c) one watt

264. A compressor having a variable clearance volume would:
    a) not be practical
    b) not affect the capacity of the system
    c) affect the capacity of the system
265. The superheat setting would be associated with:
    a) an automatic expansion valve
    b) a thermostatic expansion valve
    c) a capillary tube
266. The liquid inlet and liquid outlet of a liquid receiver should be:
    a) provided with king valves
    b) as close together as possible
    c) as far apart as practical
267. Of the refrigerants listed below, which would be permissible in a direct system for air conditioning purposes:
    a) ammonia
    b) Freon
    c) sulphur dioxide
268. A system using a centrifugal compressor would more likely operate at:
    a) high pressure
    b) low pressure
    c) medium pressure
269. The maximum allowable working pressure for which a specific part of a system is designed is called the:
    a) factor safety
    b) bursting pressure
    c) design working pressure
270. To take advantage of cheaper power rates, a system that operates only 50% of each 24-hour period should operate:
    a) condensing
    b) in the daytime
    c) at night
271. Other conditions being equal, the moisture-holding capacity of air is greater when the air temperature is:
    a) 60°
    b) 32°
    c) 120°
272. Lubricating oil is more miscible in:
    a) carbon dioxide
    b) ammonia
    c) Freon
273. The time required for ice-making may be:
    a) increased by lowering the suction pressure
    b) decreased by lowering the suction pressure
    c) decreased by a rise in suction pressure

274. The area of a circle equals:
    a) d^2 x 3.1416
    b) d x 0.7854
    c) d^2 x 0.7854
275. Upon entering the condenser, the hot gas discharge first gives up its:
    a) specific heat
    b) latent heat
    c) superheat
276. Which of the following pressure relief devices (when functioning) will cause the greatest loss of refrigerant?
    a) rupture disc
    b) relief valve connected to the receiver
    c) relief valve connected to the compressor
277. To be consistent with safety, a refrigeration operator should operate methyl chloride systems much the same as:
    a) Freon 11
    b) carbon dioxide
    c) ammonia
278. Water vapor or moisture may be removed from air by passing the air:
    a) over cooling coils
    b) over heating coils
    c) through a steam spray
279. Other conditions remaining equal, lower refrigeration temperatures require a:
    a) smaller compressor capacity
    b) larger compressor capacity
    c) higher suction pressure
280. To convert Centigrade degrees into Fahrenheit degrees:
    a) °F = 5/9 x °C + 32
    b) °F = 9/5 x °C + 32
    c) °F = 9/5 x °C - 32
281. Compressors equipped with safety heads:
    a) do not have discharge valves
    b) do not have discharge valves, as the safety heads serve this purpose
    c) have discharge valves in the safety head
282. As the temperature of lubricating oil carried over in the evaporator decreases, its viscosity will:
    a) remain unchanged
    b) decrease
    c) increase
283. A thin film of calcium carbonate, if deposited uniformly over the water side surfaces of a condenser, will:
    a) cause an acid condition in the water
    b) retard corrosion
    c) accelerate corrosion

284. The drying agent used in refrigerant driers would most likely be:
    a) sodium chloride
    b) calcium chloride
    c) silica gel
285. There is less cycling of the compressor in:
    a) an absorption system
    b) a direct system
    c) an indirect system
286. A sensation of dampness in an office or other work area is due to:
    a) high relative humidity
    b) low relative humidity
    c) high ambient temperature
287. As the pressure imposed on a refrigerant increases, the latent heat of vaporization:
    a) remains constant
    b) decreases
    c) increases
288. The stuffing box for the enclosed vertical compressor is subject to:
    a) suction pressures
    b) discharge pressures
    c) either suction or discharge pressures
289. Of the metering expansion valves listed below, which is the best for use on multiple systems?
    a) low-side float
    b) capillary tube
    c) high-side float
290. When using the same cooling water for the condenser and compressor, the automatic cooling water valve should be located at the:
    a) inlet to the compressor
    b) inlet to the condenser
    c) outlet from the condenser
291. As the relative humidity decreases, the difference in the readings of the wet and dry bulb thermometers:
    a) increases
    b) decreases
    c) remains uniform
292. Due to solubility, oil of a higher viscosity is required in systems using:
    a) Freon
    b) carbon dioxide
    c) ammonia
293. Proper maintenance of leather belts includes periodically treating with:
    a) shellac
    b) lubricating oil
    c) neatsfoot oil

294. A gauge pressure of 35 psi is equal to an absolute pressure of:
   a) 49.7 psia
   b) 20.3 psia
   c) 35 psia
295. Corrosion of the water side of a condenser is due to the presence of:
   a) scale
   b) dissolved oxygen and carbon dioxide
   c) soda ash
296. Which type of copper is not permitted in the refrigeration system:
   a) type K
   b) type L
   c) type M
297. What is the largest size soft annealed copper tubing which may be used for refrigerant piping erected on the premises?
   a) 1" OD
   b) 1-1/2" OD
   c) 1-3/8" OD
298. The freezing point of a refrigerant should be well:
   a) below the lowest evaporating temperature of the system
   b) above the highest evaporating temperature of the system
   c) above the lowest evaporating temperature of the system
299. The ambient temperature of a refrigeration compressor is the:
   a) air around the compressor
   b) temperature of the discharge vapor
   c) inlet temperature of the compressor cooling water
300. Surging in a centrifugal compressor is more apt to occur at:
   a) heavy load periods
   b) light load periods
   c) periods where non-condensable gases are present
301. A refrigerant drier will contain more moisture when:
   a) cold
   b) warm
   c) hot
302. Removable heads for cleaning and repairing condenser tubes are provided for:
   a) evaporative condensers
   b) vertical shell and tube condensers
   c) horizontal shell and tube condensers
303. The fusible plug on a receiver should be located:
   a) above the liquid line
   b) below the liquid line
   c) either above or below the liquid line

304. When a refrigerant approaches a perfect gas:
    a) its moisture content increases
    b) it becomes harder to condense
    c) it becomes easier to condense
305. When starting a motor-driven centrifugal compressor, the suction damper (when provided) should be:
    a) closed
    b) opened completely
    c) locked in the open position
306. The temperature at which particles of wax in an oil-refrigerant mixture can easily be seen by the naked eye is called the:
    a) viscosity
    b) floc point
    c) pour point
307. Increasing the superheat setting of the thermostatic expansion valve will:
    a) increase the refrigerating capacity of the unit
    b) not change the refrigerating capacity of the unit
    c) decrease the refrigerating capacity of the unit
308. Sulphur dioxide is:
    a) toxic only
    b) flammable only
    c) both toxic and flammable
309. With Freon or methyl chloride, the viscosity of an oil specified should be:
    a) lower than that used for other refrigerants
    b) higher than that used for other refrigerants
    c) the same as that used for other refrigerants
310. Calcium chloride and lithium bromide are classed as:
    a) desaturators
    b) absorbents
    c) adsorbents
311. The discharge piping from a pump-out compressor should be connected to:
    a) atmosphere
    b) the condenser
    c) the condenser as well as the atmosphere
312. Frost formation on the suction line is most generally caused by:
    a) liquid carryover from the evaporator
    b) a cold, dry atmosphere
    c) insulation on the suction line
313. The water flowing in a vertical shell and tube condenser makes a:
    a) single pass
    b) double pass
    c) triple pass

314. A system using Freon refrigerants must be vented to the outside of the building when the system:
   a) contains over 100 lbs of refrigerant
   b) contains over 50 lbs of refrigerant
   c) is over 15 tons
315. Where a compressor operates with two suction pressures, the:
   a) lowest suction pressure enters the compressor first
   b) highest suction pressure enters the compressor first
   c) compressor must be the centrifugal type
316. When a pressure relief valve and a rupture disc are used in series:
   a) the rupture disc is placed closest to the vessel protected
   b) the relief valve is placed closest to the vessel protected
   c) either may be placed closest to the vessel protected
317. Excess carbon dioxide, if inhaled, will cause:
   a) a clearness of the head
   b) a decrease in the respiration rate
   c) an increase in the respiration rate
318. Liquid refrigerant injected into the suction line ahead of the compressor will:
   a) increase the operating temperature of the compressor
   b) not affect the operating temperature of the compressor
   c) reduce the operating temperature of the compressor
319. Ethylene is classed as a:
   a) Group I refrigerant
   b) Group II refrigerant
   c) Group III refrigerant
320. A properly installed heat exchanger will:
   a) decrease the flash gas through the compressor valves
   b) decrease flash gas at the expansion valve
   c) increase flash gas at the expansion valve
321. The proper separation of liquid and vapor is best accomplished in:
   a) an atmospheric condenser
   b) a shell and tube condenser
   c) a double pipe condenser
322. With a two temperature system, the solenoid valve is generally located at the:
   a) liquid line to lower temperature box
   b) liquid line to higher temperature box
   c) suction line of high temperature box
323. The solubility of water in Freon refrigerants is:
   a) not effected by temperature changes
   b) increased by an increase in temperature
   c) decreased by an increase in temperature
324. Acrolein is used as:
   a) an aid to better oil distribution
   b) a moisture eliminator

c) a leak warning agent
325. A thermodynamic process during which no heat is extracted or added to the system is referred to as:
   a) the point of no heat transfer
   b) an isothermal process
   c) an adiabatic process
326. Dual pressure relief devices are required when the pressure vessel is:
   a) 10 ft^3 or more
   b) 3 ft^3 or more
   c) over 10 feet long
327. Medium and large compressors are rated in:
   a) horsepower
   b) Btu/hour
   c) operating pressure
328. The common unit of heat energy is the:
   a) Btu
   b) Fahrenheit degree
   c) ft/lb
329. Which of the following compressors is best suited for intermittent (cycling) operation?
   a) vertical double-acting
   b) centrifugal
   c) vertical single-acting
330. For use in ultra-low temperature systems, the most practical type of thermostatic expansion valve would be one employing a:
   a) gas charge
   b) liquid charge
   c) cross charge
331. Density and latent heat being considered, less condenser is required for:
   a) Freon
   b) methyl chloride
   c) ammonia
332. When shut down for the winter, the refrigerant in an ice-making plant should be stored in the:
   a) cooling coils
   b) accumulator
   c) condenser or receiver
333. An indicator card having a suction line that starts below the normal suction pressure line and meets the normal suction line before completion of the stroke indicates that the:
   a) suction valve spring may be too stiff
   b) suction valve spring may be too weak or loose
   c) compressor discharge is not high enough

334. In six hours, a one-ton compressor will handle a load of:
    a) 288,000 Btu
    b) 144,000 Btu
    c) 72,000 Btu
335. The booster compressor discharges into the:
    a) suction side of the main compressor
    b) discharge side of the main compressor
    c) inlet side of the condenser
336. When Freon 12 is sometimes added to the Freon 22 charge in the system, the amount is approximately 3% to:
    a) make one system more workable by eliminating free oil
    b) makes leaks less expensive
    c) be able to test for leaks with a halide torch
337. Capacity of a centrifugal compressor is often governed by:
    a) low-side float control
    b) changing the condenser pressure
    c) the use of clearance pockets
338. Rusting and corrosion of metal parts in a refrigeration system using Freons is greater with:
    a) water dissolved in refrigerant
    b) water dissolved in oil in the system
    c) undissolved water
339. Where is the oil separator located in the system?
    a) between the condenser and the liquid line
    b) between the receiver and the liquid line
    c) between the compressor and the condenser
340. With water-cooled condensers, the requirement per ton of refrigeration is :
    a) 1 to 5 gal/min
    b) 5 to 10 gal/min
    c) 10 to 15 gal/min
341. Automatic expansion valves function best when used in conjunction with a:
    a) temperature control switch (thermostat)
    b) back pressure control switch
    c) high pressure cut-out switch
342. Of the substances listed below, which is considered a chlorinated refrigerant?
    a) carbon dioxide
    b) ammonia
    c) Freon
343. Clear water:
    a) will always make clear ice
    b) is considered hard water
    c) does not necessarily make clear ice

344. The lower the suction pressure, the:
   a) less the cost of power to operate the system
   b) less volume of vapor the compressor must handle
   c) greater volume of vapor the compressor must handle
345. When a vapor is in equilibrium and in contact with the liquid from which it was formed, it is referred to as a:
   a) wet vapor
   b) superheated vapor
   c) saturated vapor
346. The main function of the water jacket surrounding the head and cylinder of a compressor is to:
   a) pre-cool the refrigerant before it enters the compressor
   b) keep the cylinder walls cool and facilitate lubrication
   c) remove superheat
347. Usually, the dry bulb (db) temperature is:
   a) the same as the wet bulb temperature
   b) lower than the wet bulb temperature
   c) higher than the wet bulb temperature
348. Compressors having no high- to low-side equalizing device for starting purposes require drives with:
   a) unloaders
   b) high starting torques
   c) low starting torques
349. The high-side float valve is ordinarily connected to:
   a) one evaporator
   b) more than one evaporator
   c) the capillary tube
350. Expansion at the expansion valve actually occurs in the:
   a) vapor
   b) liquid
   c) metal structure of the valve
351. Vertical shell and tube condensers may be located:
   a) only on the inside of buildings
   b) either inside or outside of buildings
   c) only on the outside of buildings
352. All conditions being equal, the pressure differential:
   a) must be greater in an indirect system
   b) will be the same for both direct and indirect systems
   c) must be greater in a direct system
353. The temperature of the medium being cooled in a unit must be:
   a) above the evaporator temperature
   b) below the evaporator temperature
   c) equal to the evaporator temperature

354. The axiom that energy can neither be created or destroyed, but can be converted from one form to another pertains to:
    a) the second law of thermodynamics
    b) the first law of thermodynamics
    c) Boyle's law
355. The larger the compressor size, the:
    a) slower its speed of rotation
    b) faster its speed of rotation
    c) slower its flywheel rpm speed
356. The rate of heat transfer in a condenser from the refrigerant to the water:
    a) decreases as the viscosity of the refrigerant increases
    b) increases as the viscosity of the refrigerant increases
    c) means an evaporative condenser is being used
357. The bellows of a high pressure cut-out switch contains:
    a) a vacuum
    b) a refrigerant
    c) air
358. If the evaporator section to which the feeler bulb is attached is completely filled with unevaporated refrigerant, the expansion valve should be:
    a) open or be opening
    b) closed or be closing
    c) open fully
359. A shell and tube condenser will operate more efficiently if it is of the:
    a) counter flow type
    b) parallel flow type
    c) series flow type
360. The function of a thermometer well is to:
    a) store the thermometer expansion fluid
    b) facilitate thermometer removal without refrigerant loss
    c) keep the thermometer bulb cold
361. A sensation of dryness in an office or other work area is due to:
    a) high relative humidity
    b) low relative humidity
    c) low ambient temperature
362. The part of the refrigeration system which adds the most heat to the refrigerant is the:
    a) compressor
    b) condenser
    c) evaporator
363. If the refrigerant leaving the expansion coils is in a superheated condition, the system is classed as:
    a) an absorbtion system
    b) a flooded system
    c) a dry expansion system

364. In a stem thermometer, the hollow space above the liquid:
   a) should be filled with air
   b) may possibly contain gas vaporized from this liquid
   c) is under a perfect vacuum
365. A refrigerant having a high boiling point requires:
   a) a low pressure or vacuum on the suction side
   b) a high pressure on the suction side
   c) use of a reciprocating type of compressor
366. A method used in ammonia systems to recover refrigerant from the lubricating oil is the use of:
   a) oil chillers
   b) oil separators
   c) oil stills
367. Indicator diagrams are taken while the compressor is:
   a) down for overhaul
   b) stopped
   c) running
368. An expansion device, which in the strict sense of the word is not a valve, is:
   a) thermostatic type
   b) low-side float
   c) capillary tube
369. An internal receiver pipe connected to the king valve should:
   a) have the open end facing upwards
   b) have the open end facing downwards
   c) not be used
370. When a moisture seal or barrier is used in connection with insulation, it should be placed:
   a) adjacent to the cold side
   b) adjacent to the warm side
   c) exactly in the center
371. A saturated vapor is a vapor:
   a) saturated with liquid droplets
   b) at a temperature corresponding to the pressure
   c) at a temperature higher than 5°F
372. During operation, the shaft seal on a Carrier Centrifugal Machine is:
   a) a metal-to-metal seal
   b) a metal-to-carbon seal
   c) an oil pressure seal
373. A field test is a test performed:
   a) in the field, to prove system tightness
   b) to prove that the system is operating properly
   c) to prove that the system is sized properly

374. The "K" factor of an insulating material defines:
   a) Btu loss
   b) moisture resistance
   c) combustible qualities
375. When a calcium chloride brine begins foaming, it might possibly indicate:
   a) lack of aeration
   b) a brine that is too strong
   c) occurrence of corrosion
376. The most perfect insulator would be:
   a) a vacuum
   b) still air
   c) cork
377. A 10 x 10 compressor refers to one:
   a) equalling 100 in² displacement volume
   b) having a 10" bore and 10" stroke
   c) of 10 horsepower and 10 tons of refrigeration
378. On an ammonia system, the scale trap is preferably located on the:
   a) suction line close to the evaporator
   b) discharge line close to the compressor
   c) suction line close to the compressor
379. The piping on the liquid lines on an ammonia system would be:
   a) schedule 20
   b) schedule 40
   c) schedule 80
380. A thermometer well should be:
   a) packed with asbestos
   b) left dry
   c) filled with oil
381. An air conditioning system serving a hospital should distribute air so that the:
   a) relatively cooler air exhausted from the operating room is recirculated to the convalescence rooms or wards
   b) air from the operating rooms is not distributed to other rooms
   c) relative humidity and temperature is carried as low as possible
382. When leak-testing a CO_2 compressor system with air pressure, it is considered good practice to:
   a) use the CO_2 compressor to build up the necessary air pressure
   b) use a lubricating oil having a low flash point
   c) obtain the necessary air pressure through a separate air pressure
383. Saturation temperature corresponding to the critical state of the substances at which the properties of the liquid and vapor are identical is known as:
   a) absolute temperature
   b) saturation temperature
   c) critical temperature

384. A safety collar is used in conjunction with a:
   a) discharge valve
   b) built-up piston
   c) suction valve
385. A sizeable leak in the float ball of a high-side float valve causes the valve to:
   a) remain closed
   b) remain open
   c) open and close intermittently
386. A higher condensing temperature will require:
   a) less quantity of water
   b) a lower condensing pressure
   c) a higher condensing pressure
387. A back pressure valve installed on a system employing a liquid chiller prevents the evaporator from becoming:
   a) too cold
   b) too warm
   c) saturated with liquid
388. Leaking piston rings and leaking valves will be indicated by:
   a) abnormally low discharge temperatures
   b) abnormally high discharge temperatures
   c) increased frost on the suction lines
389. Which of the following would not be classed as a Group II refrigerant?
   a) ammonia
   b) sulphur dioxide
   c) Butane
390. What keeps a flat belt from slipping off a pulley?
   a) pulley is made concave
   b) static electricity developed by the belt
   c) weight of the belt
391. The device that compresses and delivers the refrigerant at a temperature which is adequately above that of the atmosphere to the unit where the heat energy is discarded is called:
   a) a heat pump
   b) a compressor
   c) an expansion valve
392. After leaving the compressor, the high pressure, high temperature, and refrigerant vapor is passed through:
   a) an evaporator
   b) a condenser
   c) a receiver
393. If a thermometer was placed at the outlet part of the compressor, it would measure only:
   a) latent heat

b) superheat
c) sensible heat
394. The proper operation of the high-side float depends chiefly on the:
    a) correct amount of refrigerant charge in the system
    b) size of the condenser
    c) capacity of the compressor
395. The high-side float controls the flow of the refrigerant:
    a) to the evaporator
    b) from the evaporator
    c) from the condenser
396. One of the causes of high discharge temperatures is:
    a) liquid ammonia in the suction side
    b) the expansion valve is open too wide
    c) accumulation of scale or dirt in the compressor cooling water jackets
397. The expansion valve of a refrigeration system needs:
    a) frequent adjustment
    b) no adjustment
    c) adjusting occasionally as indicated by frost on frost pipe
398. Scale traps on direct expansion systems are:
    a) underneath the scales
    b) on suction side of the compressor
    c) on discharge side of the compressor
399. The refrigerants ammonia and carbon dioxide should be stored:
    a) in the direct rays of the sun
    b) near some source of heat
    c) away from any source of heat
400. The high-side of a refrigeration system is:
    a) in the highest side of the building
    b) near some source of heat
    c) away from any source of heat
401. To recharge weak brine solution:
    a) add concentrated brine solution at inlet of system
    b) dump calcium chloride into the tank
    c) suspend a sack containing calcium chloride at the outlet of the system
402. When ice melts, its temperature:
    a) increases slightly
    b) increases by degrees
    c) remains constant
403. Assuming that a compressor is in good condition, what limits the degree of vacuum that it will develop?
    a) speed of compressor
    b) kind of refrigerant used
    c) clearance volume of the compressor

404. Which refrigerant can be tested for leaks by using a halide leak detector?
    a) R-12
    b) R-717
    c) R-744
405. What is the micron equivalent of 1 inch of vacuum?
    a)   500
    b) 2,400
    c) 25,400
406. Condensed refrigerant travels from the condenser to the expansion valve through the:
    a) suction line
    b) discharge line
    c) liquid line
407. A hissing noise at the expansion valve is a good indication that:
    a) liquid is passing through
    b) gas is passing through
    c) nothing is passing through
408. The Freon group of refrigerants has:
    a) an odor similar to sulphur dioxide
    b) a very pungent odor
    c) a slightly sweet odor
409. The purpose of a vent or equalizing line between a condenser and receiver is to insure flow:
    a) to the atmosphere in the event of an emergency
    b) from the receiver to the evaporator
    c) from the condenser to the receiver
410. Heat transfer by conduction is faster in a:
    a) superheated vapor
    b) saturated vapor
    c) liquid
411. Which refrigerant listed is most suitable for ice-making in a hospital?
    a) ammonia
    b) carbon dioxide
    c) sulphur dioxide
412. The brine generally used in ice-making plants is:
    a) sodium chloride
    b) calcium chloride
    c) barium chloride
413. Pressure relief valves protecting compressors from overpressure are usually vented to:
    a) an absorbing solution
    b) the low-side
    c) atmosphere

414. Which of the following refrigerants is lighter than air?
    a) Freon
    b) carbon dioxide
    c) ammonia
415. The temperature of the liquid line should normally be:
    a) the same as the condensing water temperature
    b) higher than the condensing water temperature
    c) lower than the condensing water temperature
416. A binary system has two:
    a) different refrigerants
    b) suction pressures
    c) discharge pressures
417. Of the refrigerants listed, the one most suitable to air-condition a theater is:
    a) Freon 11
    b) ammonia
    c) propane
418. An automatic expansion valve:
    a) properly maintains a constant pressure and temperature in the evaporator
    b) responds well to large changes in heat load
    c) works well with systems requiring several temperatures
419. Freon 12 is composed of:
    a) hydrogen and nitrogen
    b) carbon and oxygen
    c) fluorine, chlorine and carbon
420. The unit used for measuring gas pressure is the:
    a) pound
    b) Btu
    c) psi
421. Which of the following cannot be used in a mechanical refrigeration system?
    a) swaged joints on the low-side
    b) flared joints
    c) 50/50 solder joints
422. A vapor that has been heated above its boiling point is referred to as a:
    a) saturated vapor
    b) superheated vapor
    c) sub-heated vapor
    d) latent heated vapor
423. The process by which a substance changes from a gaseous to a liquid state is referred to as:
    a) evaporation
    b) absorption
    c) vaporization
    d) condensation

424. The device in which the refrigerant is vaporized is called the:
   a) compressor
   b) vaporizer
   c) evaporator
   d) condenser
425. The device that regulates the flow of liquid refrigerant to the cooling unit so that the evaporator is maintained neatly full of liquid refrigerant that does not enter the suction line is called the:
   a) compressor
   b) receiver
   c) condenser
   d) expansion valve
426. The refrigerant storage tank, located below the condenser to hold a surplus of refrigerant, is called the:
   a) receiver
   b) evaporator
   c) accumulator
   d) well
427. Temperature is measured accurately with a:
   a) manometer
   b) thermometer
   c) barometer
   d) thermostat
428. The odor of ammonia is:
   a) pungent
   b) sweet
   c) sour
   d) no smell
429. Carbon dioxide refrigerant:
   a) is pungent
   b) is odorless
   c) smells like phosgen gas
   d) sweet
430. The average weight of ice cakes manufactured by the large plants is:
   a) 100 lbs
   b) 300 lbs
   c) 700 lbs
   d) 900 lbs
431. The latent heat of evaporation is highest for:
   a) water
   b) ammonia
   c) carbon dioxide
   d) R-500

432. Ammonia in the presence of sulphur vapor produces a:
    a) white smoke
    b) black smoke
    c) green flame
    d) blue flame
433. The temperature variation of the brine in an ice field should not exceed:
    a) 2°F to 5°F
    b) 0°F to 2°F
    c) 5°F to 10°F
    d) 10°F to 15°F
434. Carbon dioxide is composed of:
    a) hydrogen and nitrogen
    b) carbon and oxygen
    c) fluorine, chloride and carbon
    d) CO_2 and NH_3
435. The electric motor that drives the compressor in home units is turned on and off by the:
    a) receiver
    b) condenser
    c) cooling units
    d) thermostat
436. After leaving the compressor, the vaporous refrigerant goes to the:
    a) receiver
    b) condenser
    c) thermostat
    d) liquid flow control
437. The device used to make the compressor in a household refrigerator is:
    a) an electric motor
    b) a generator
    c) a heat engine
    d) an air motor
438. Mercury is used in thermometers primarily because it:
    a) expands uniformly
    b) vaporizes at very high temperatures
    c) is cheaper than most other suitable liquids
    d) never freezes
439. The operation of every type of motor is based on the:
    a) principle of induction
    b) frequency of the power supply
    c) principle of attraction
    d) magnetic and electric reactions
440. The unit used for force is the:
    a) pound

b) pound of feet
c) pounds per square inch
d) horsepower

441. The unit for measuring mechanical power is the:
    a) foot-pound
    b) pound
    c) Btu
    d) horsepower

442. The unit used for measuring electrical power is the:
    a) ampere
    b) horsepower
    c) watt
    d) Btu

443. One mechanical horsepower is defined as the work done at the rate of:
    a) 778 Btu per minute
    b) 550 ft-lbs/second
    c) 3,300 ft-lbs/second
    d) 3,300 ft-lbs/minute

444. For protection against overloading single-phase induction motors, it is customary to use:
    a) thermal relays
    b) fuses
    c) compensators
    d) fusetrons

445. Electrical branch circuit conductors supplying an individual motor must have a carrying capacity that exceeds the motor's full load current rating by:
    a) 25%
    b) 50%
    c) 125%
    d) 150%

446. The rating of the fuses or overcurrent devices used to protect motors must not exceed the current capacity of the conductors by more than:
    a) 5%
    b) 10%
    c) 15%
    d) 25%

447. The latent heat of vaporization of sulphur dioxide is:
    a) 144 Btu
    b) 168 Btu
    c) 746 Btu
    d) 970 Btu

448. The latent heat of vaporization of water is:
    a) 144 Btu

b) 168 Btu
c) 746 Btu
d) 970 Btu

449. The specific heat of water is:
   a) 1.000
   b) 0.648
   c) 0.463
   d) 0.481

450. The silver-coated contacts in temperature and motor controls should be cleaned with:
   a) a crocus cloth
   b) an emery cloth
   c) extra fine sandpaper
   d) an extra fine, smooth-cut file

451. One British thermal unit (Btu) is the amount of heat required to raise the temperature of:
   a) one pound of water 1°F
   b) one gallon of water 1°F
   c) one cubic foot of water 1°F
   d) one pound of water 1°C

452. The process by which a substance is changed from a solid to a liquid state is referred to as:
   a) melting
   b) dissolving
   c) vaporization
   d) disintegration of the atom

453. To melt one ton of ice (2000 lbs) requires:
   a) 144 Btu
   b) 2,000 Btu
   c) 1,200 Btu
   d) 288,000 Btu

454. The heat given up by a substance during the freezing process is called:
   a) latent heat of fusion
   b) latent heat of vaporization
   c) evolution
   d) condensation

455. The frost on the evaporating coils:
   a) acts as an insulator
   b) decreases the heat pressure
   c) increases the temperature range
   d) increases overall efficiency

456. The differential adjustment in the temperature control unit controls the:
   a) defrosting interval

b) cycling interval
c) amount of refrigerant that can enter the receiver
d) cabinet temperature range

457. The high-side of a system usually has a purge valve to:
    a) remove oil from the system
    b) add oil to the system
    c) add refrigerant to the system
    d) remove air from the system

458. The amount that flows into the high-side float chamber is determined primarily by the:
    a) evaporating unit capacity
    b) condensing unit capacity
    c) low-side float capacity
    d) size of the compressor

459. The refrigerant considered the safest for use in domestic refrigeration is:
    a) sulphurdioxide
    b) Freon 12
    c) methyl chloride
    d) ammonia

460. The refrigerant with which lubricating oil mixes best is:
    a) Freon 12
    b) SO_2
    c) ethyl chloride
    d) methyl chloride

461. A device that transfers a gaseous substance from a container of low pressure to one of high pressure is called:
    a) an evaporator
    b) a receiver
    c) a high-side float
    d) a compressor

462. A pressure-reducing valve that is placed in the liquid line just before it enters the cooling unit is called:
    a) a check valve
    b) a low-pressure valve
    c) a high-side float valve
    d) an expansion valve

463. The length of the refrigerator's on/off cycles depends upon:
    a) the setting of the thermostat
    b) how much or how little refrigerant is used
    c) the type of temperature control used
    d) the type of refrigerant used

464. The lift of the compressors suction and discharge valve should be:
    a) high
    b) low

c) by hand
d) heat

465. The best conduction of heat would take place through a:
    a) non-metal object
    b) ferrous metal
    c) perfect vacuum
    d) liquid

466. Of the metals listed below, Freon 11 has the greatest corrosive effect on:
    a) copper
    b) steel
    c) magnesium
    d) tin

467. When cleaning the tubes of a vertical shell and tube condenser by mechanical means, it is considered good practice to:
    a) keep the water flowing through the tubes
    b) stop the flow of water through the tubes
    c) use ammonia as the cleaning agent
    d) use SO_2 as the cleaning agent

468. If a small part of the system required a much lower temperature than the remainder of the system, it would be best to utilize a:
    a) centrifugal compressor
    b) booster compressor
    c) pump-out compressor
    d) pump-in compressor

469. An Ammonia system generally utilizes a:
    a) reciprocating compressor
    b) rotary compressor
    c) centrifugal compressor
    d) gear compressor

470. Of the metals listed below, NH_3 has the greatest corrosive effect on:
    a) steel
    b) copper
    c) aluminum
    d) lead

471. A trunk-type piston, compared to the ordinary type, is:
    a) the same length
    b) much shorter
    c) much longer
    d) made of lead

472. To detect carbon dioxide leaks, it is considered good practice to use:
    a) oil of peppermint
    b) a sulphur stick
    c) an ammonia swab
    d) R-11

473. Ammonia is composed of:
    a) hydrogen and nitrogen
    b) carbon and oxygen
    c) fluorine, chloride and carbon
    d) SO_2 and CCl_2F_2
474. Vertical single-acting compressors normally discharge gas:
    a) at the top of the cylinder
    b) at the bottom of the cylinder
    c) through the safety head
    d) at the side of the unit
475. Temperature is a term used to designate the:
    a) amount of energy in the substance
    b) intensity of energy in the substance
    c) amount of heat in the substance
    d) intensity of heat in the substance
476. The unit of measurement for temperature is:
    a) Btu
    b) ohm
    c) calorie
    d) Fahrenheit
477. One Btu of work is equal to:
    a) one horsepower of work
    b) 42.42 horsepower of work
    c) one ft-lb of work
    d) 778 ft-lbs of work
478. The temperature at which a substance changes from liquid to vapor is:
    a) boiling point
    b) condensing point
    c) changing point
    d) latent heat point
479. Heat that increases temperature of a refrigerant without causing a change in state is called:
    a) sensible heat
    b) latent heat
    c) specific heat
    d) vaporizing heat
480. The quantity of heat in a refrigerator may be measured by a unit called the:
    a) Fahrenheit
    b) foot-pound
    c) calorie/in^2
    d) Btu
481. If it were possible to remove all the heat from a refrigerator, its temperature would be:
    a) 32°F

b) 0°F
c) -40°F
d) -460°F

482. The purpose of the electric motor is to:
    a) increase the energy put into it
    b) generate power
    c) change mechanical energy into electrical energy
    d) change electrical energy into mechanical energy
483. In determining the horsepower of a motor needed to operate a particular refrigerating unit, main consideration should be given to:
    a) the temperature rise of the motor
    b) operating characteristics of the motor
    c) the slip of the motor
    d) how fast it will stop
484. The type of single-phase motor that has the best starting torque is the:
    a) capacitor type
    b) shaded coil type
    c) shaded pole type
    d) split-phase type
485. Excessive motor vibration may be caused by:
    a) overloading the motor
    b) light loading of the motor
    c) the pulley being too tight
    d) worn armature shaft bearings
486. Unless it's a type approved for more, the maximum number of electrical conductors that may be placed under a screw type terminal is:
    a) one
    b) two
    c) three
    d) four
487. In a refrigerator, the food is preserved because the unit:
    a) puts cold into storage space
    b) removes heat from storage space
    c) allows no heat to get into the storage space
    d) will not permit cold to get out of the storage space
488. Heat is carried to the refrigeration cooling unit by:
    a) radiation
    b) convection
    c) conduction
    d) heat coils
489. The device used to maintain the temperature inside a refrigerator within a certain limit is called a:
    a) thermometer
    b) pyrometer

c) governor
d) thermostat

490. The amount of heat that a substance contains is measured with an instrument called a:
    a) thermometer
    b) potentiometer
    c) calorimeter
    d) pyrometer

491. The latent heat of vaporization of Freon 12 at 5°F is:
    a) 68.2 Btu/lb
    b) 169 Btu/lb
    c) 565 Btu/lb
    d) 970 Btu/lb

492. White ice is the result of:
    a) too much ammonia
    b) excessive oxygen
    c) insufficient air agitation

493. Refrigeration is a process of:
    a) making ice
    b) preserving food
    c) extracting heat
    d) adding heat

494. An efficient test for sulphur dioxide leaks is:
    a) sulphur sticks
    b) litmas paper
    c) ammonia swab
    d) R-502

495. The color code for R-500 would be:
    a) green
    b) yellow
    c) white
    d) orange

496. The color code for R-502 would be:
    a) green
    b) yellow
    c) purple
    d) orange

497. The chemical name for R-717 would be:
    a) methyl chloride
    b) sulphur dioxide
    c) ammonia
    d) water

498. The chemical name for R-718 is:
   a) methyl chloride
   b) sulphur dioxide
   c) ammonia
   d) water
499. The chemical name for R-764 is:
   a) methyl chloride
   b) sulphur dioxide
   c) carbon dioxide
   d) ammonia
500. The chemical name for R-744 is:
   a) methyl chloride
   b) sulphur dioxide
   c) carbon dioxide
   d) ammonia
501. The chemical name for R-40 would be:
   a) methyl chloride
   b) sulphur dioxide
   c) carbon dioxide
   d) ammonia
502. The chemical name for R-12 would be:
   a) dichlorodifluoromethane
   b) chlorodifluoromethane
   c) trichloromonofluoromethane
   d) methyl chloride
503. The chemical name for R-22 would be:
   a) dichlorodifluoromethane
   b) chlorodifluoromethane
   c) trichloromonofluoromethane
   d) methyl chloride
504. The chemical formula for R-717 would be:
   a) SO_2
   b) NH_3
   c) CO_2
   d) H_2O
505. The chemical formula for R-744 would be:
   a) SO_2
   b) NH_3
   c) CO_2
   d) H_2O
506. The chemical formula for R-764 would be:
   a) SO_2
   b) NH_3

c) CO_2
d) H_2O

507. The chemical formula for R-718 would be:
    a) SO_2
    b) NH_3
    c) CO_2
    d) H_2O

508. The boiling point for R-764 would be:
    a) -109
    b) 14
    c) -28
    d) 212

509. The boiling point for R-718 would be:
    a) -109
    b) 14
    c) -28
    d) 212

510. The boiling point for R-12 would be:
    a) 14
    b) -21.6
    c) -41.4
    d) -28

511. The boiling point for R-22 would be:
    a) 14
    b) -21.6
    c) -41.4
    d) -28

512. The boiling point for R-500 would be:
    a) 14
    b) -21.6
    c) -41.4
    d) -28

513. The chemical formula for R-290 would be:
    a) CO_2
    b) SO_2
    c) C_2H_6
    d) C_3H_8

514. The chemical formula for R-170 would be:
    a) CO_2
    b) SO_2
    c) C_2H_6
    d) C_3H_3

515. The chemical formula for R-600 would be:
   a) C_4H_{10}
   b) C_3H_3
   c) C_2H_6
   d) CCl_3F
516. The boiling point for R-717 would be:
   a) 14
   b) -28
   c) -50
   d) 212
517. The boiling point for R-744 would be:
   a) 14
   b) 212
   c) -109
   d) -50
518. The chemical formula for R-40 would be:
   a) SO_2
   b) CO_2
   c) CH_3Cl
   d) $CHClF_2$
519. The chemical formula for R-12 would be:
   a) $CHClF_2$
   b) CCl_2F_2
   c) CH_3Cl
   d) CH_2L_2F
520. The chemical formula for R-22 would be:
   a) $CHClF_2$
   b) CCl_2F_2
   c) CH_3Cl
   d) CH_2L_2F
521. The chemical formula for R-11 would be:
   a) C_3H_3
   b) C_4HLO
   c) CCl_3F
   d) CH_2L_2F
522. The chemical formula for R-30 would be:
   a) CH_2Cl_2
   b) C_4HLO
   c) C_3H_3
   d) C_2H_6
523. The boiling point for R-502 would be:
   a) -50
   b) -41.4

c) -28
d) 14

524. The color code for R-717 would be:
    a) silver
    b) green
    c) white
    d) orange
525. The color code for R-764 would be:
    a) silver
    b) gray
    c) black
    d) green
526. The color code for R-744 would be:
    a) silver
    b) gray
    c) black
    d) orange
527. The color code for R-12 would be:
    a) green
    b) yellow
    c) white
    d) orange
528. The color code for R-22 would be:
    a) green
    b) yellow
    c) white
    d) orange
529. The color code for R-11 would be:
    a) green
    b) yellow
    c) white
    d) orange
530. The liquid refrigerant drops to evaporate temperature:
    a) in the liquid line
    b) in the evaporator
    c) at the expansion valve
531. The condenser most likely to be found on the roof is:
    a) a shell and tube vertical
    b) an air-cooled type
    c) an atmospheric
532. With a starved evaporator, the suction pressure is:
    a) high
    b) low
    c) the amount of refrigerant in the evaporator has no bearing on pressure

533. Leak test an ammonia system with:
    a) a sulphur stick
    b) a halide torch
    c) by placing system under vacuum
534. Leak test a methyl chloride system with:
    a) a sulphur stick
    b) a halide torch
    c) by placing the system under vacuum
535. A dry evaporator:
    a) has no refrigerant in it and is fed with a TXV
    b) is 1/4 to 1/3 full of liquid and is fed with a low-side float
    c) is about 1/3 full of liquid refrigerant and may be fed with a high-side float
536. An accumulator:
    a) keeps coils flooded and separates liquid from a vapor
    b) acts as a separator only
    c) acts as a storage vessel for excess refrigerant
537. Baffles are inserted with the evaporator coils to:
    a) increase the air circulation over the coils
    b) decrease the air circulation over the coils
    c) prevent stratification
538. A CO_2 system in which the water enters the condenser at 93° will:
    a) not operate on the liquid cycle
    b) not operate as a refrigerant system
    c) operate but will not be efficient
539. Ethelene is in refrigerant:
    a) Group I
    b) Group II
    c) Group III
540. Which would use a higher velocity oil?
    a) Freon
    b) NH_3
    c) CO_2
541. When you turn a TXV valve clockwise, you:
    a) increase the superheat
    b) decrease the superheat
    c) decrease the pressure and increase the temperature
542. A spherical piston head is used to:
    a) increase volumetric efficiency and strengthen the piston
    b) ease the installation of piston rings
    c) keep down the turbulence of the compressed gas
543. The solubility of water in Freon is greater at:
    a) low temperature
    b) high temperature
    c) neither

544. What would cause brine to foam?
    a) weak brine and air
    b) an NH_3 leak
    c) too much agitation
545. Saturated NH_3 has:
    a) few liquid droplets
    b) no liquid particles
    c) some liquid and some gas
546. If more pressure is applied on refrigerant, its boiling point:
    a) increases
    b) decreases
    c) stays the same
547. Atmospheric condensers use:
    a) less water
    b) more water
    c) no water
548. A vertical two-piston compressor is:
    a) single-stage
    b) two-stage
    c) three-stage
549. A gas drier in an F-12 system would most likely have:
    a) silica gel
    b) activated alumins
    c) calcium sulphate
550. A Freon 22 system with 3% Freon 12 would have the Freon 12 added to:
    a) reduce costs through leaks
    b) enable you to find leaks with a halide torch
    c) help circulate the oil through the system
551. When no heat is added to a process, it is called:
    a) an adiabatic process
    b) an isothermal process
    c) a superheat process
552. When water freezes, it:
    a) expands
    b) contracts
    c) does not change in volume
553. At a very low temperature, use:
    a) NH_3
    b) Freon 12
    c) Freon 11
554. The latent heat of fusion of water is:
    a) 32°
    b) 144 Btu
    c) 0°

555. The best refrigerant for -110° is:
   a) Freon
   b) NH_3
   c) CO_2
556. Which has a lower moisture content?
   a) 55°db, 36°wb
   b) 35°db, 50°wb
   c) 50°db, 33°wb
557. Which high-pressure refrigerant gas entering the condenser is given up first?
   a) latent heat
   b) sensible heat
   c) superheat
558. On a centrifugal compressor, the pressure on the seal is:
   a) outside the seal
   b) inside the seal
   c) both inside and outside the seal
559. A thin line of bromide on the pipes would:
   a) leave scale deposits
   b) corrode the metal
   c) leave lime deposits
560. An indicator chart is hooked to a:
   a) compressor
   b) suction line
   c) TXV
561. The pressure limiting device (safety):
   a) may be field set
   b) may not be field set
   c) may be field set by a refrigeration contractor only
562. What temperature is best for making ice?
   a) 0°F
   b) 16°F
   c) 28°F
563. Air is injected into freezing water to:
   a) purify the ice
   b) help freeze the water
   c) agitate the ice water
564. An oil separator is located in the:
   a) suction line
   b) liquid line
   c) discharge lines
565. The bearing surface of a bellows type of shaft seal:
   a) rubs the shaft
   b) is separated by a thin film of oil
   c) rotates with the shaft

566. The compressor more likely to use a bellows-type shaft is:
    a) horizontal
    b) vertical single-acting
    c) rotary
567. With a back pressure control, the refrigeration unit would:
    a) start on pressure drop, stop on pressure rise
    b) stop on pressure rise, start on pressure drop
    c) stop on pressure drop, start on pressure rise
568. Which relieves more pressure?
    a) relief valve
    b) rupture disc
    c) fusible plug
569. The shaft seal that is subject to the highest pressure on the compressor is:
    a) horizontal double-acting
    b) vertical single-acting
    c) rotary compressor
570. A low-pressure oil cut-out stops the compressor on:
    a) high oil pressure
    b) low oil pressure
    c) low oil level
571. In starting a CO_2 system:
    a) the suction and discharge valves should be open
    b) the discharge valves should be open
    c) the H_2O should be shut off to the condenser
572. The vapor passing through the discharge valve of the compressor is:
    a) slightly above the condensing pressure
    b) slightly below the condensing pressure
    c) at the condensing pressure
573. Leaking discharge valves on a compressor will cause:
    a) high head pressure
    b) the unit will short cycle
    c) low suction pressure
574. A bellows type shaft seal:
    a) moves diagonally only
    b) rotates with the shaft
    c) is fixed and stationary
575. Piston rings are sometimes bored on an eccentric:
    a) for longer life
    b) to make the rings heavier
    c) to make the rings expand against the cylinder wall
576. A drum (cylinder) of refrigerant is:
    a) filled to capacity
    b) partially filled (80%)
    c) filled full because it is practical, it is protected by the fusible plug

577. The container in which refrigerant is stored must conform with:
    a) ASME code
    b) the state safety code
    c) Interstate Commerce Commission
578. An operator must keep the equipment and engine room clean, as required by:
    a) state code
    b) ASME B-9 code
    c) neither a or b
579. The type of pipe used on an ammonia system for the low-side:
    a) schedule 20
    b) schedule 40
    c) schedule 60
580. In institutional occupancy, refrigerant piping:
    a) may pass between the floors
    b) may not pass between the floors
    c) either a or b
581. Oil problems are more common with:
    a) Freon 12
    b) Freon 22
    c) NH_3
582. The high-pressure cut-out is installed:
    a) before the first assembly
    b) after the first assembly
    c) without a stop valve
583. On a centrifugal compressor, the shaft seal is:
    a) metal-to-metal
    b) metal-to-carbon
    c) an oil seal
584. Good insulation will have:
    a) a high K factor
    b) a low K factor
    c) any K factor
585. Evaporator condensers are placed on the roof because:
    a) there is more space
    b) the static pressure is greater
    c) the evaporation is better
586. The king valve of a liquid receiver is between the:
    a) condenser and the receiver
    b) receiver and the expansion valve
    c) accumulator and the compressor
587. The opening of the king valve faces (inlet):
    a) up from the receiver
    b) drain from the receiver
    c) cross the liquid receiver

588. Water enters a shell and tube condenser at:
   a) the top
   b) the bottom
   c) neither top or bottom
589. In a cabinet, the vapor barrier is:
   a) next to the inner wall
   b) next to the outer walls
   c) midway between the two walls
590. High pressure in the evaporator is such that when the refrigerant boils at 15°F:
   a) all the evaporator will be at this temperature
   b) none of the evaporator will be at this temperature
   c) some of the evaporator will be above and some below this temperature
591. A low boiling temperature of the refrigerant in the evaporator is obtained by:
   a) a vacuum
   b) raising the pressure
   c) slowing down the compressor
592. What is the maximum (plus or minus) voltage variation allowable at an outlet?
   a) 10%
   b) 15%
   c) 20%
   d) 25%
593. NH_3 vapor at saturation is a:
   a) wet vapor with liquid droplets in it
   b) dry vapor
   c) dry vapor that is exactly at evaporating temperature
594. Air passing through a 40° evaporator:
   a) increases in relative humidity
   b) decreases in relative humidity
   c) reaches its dew point
595. The true compression curve is likely to be closer to the:
   a) power curve
   b) isothermical curve
   c) adiabatic curve
596. Brine with the highest specific heat is:
   a) NaCl (sodium chloride)
   b) CaCl (calcium chloride)
   c) H_2O
597. For human comfort, a great deal of humidity in the air would create:
   a) a light load on the compressor
   b) an average load on the compressor
   c) a heavy load on the compressor
598. The freezing point of a refrigerant, in regard to its evaporating temperature, must be:

a) higher
b) lower
c) either higher or lower

599. Specific gravity is the:
   a) ratio of the head pressure divided by the suction pressure
   b) weight of the refrigerant
   c) ratio of its weight to the weight of the same volume of H_2O

600. If a room was saturated with CO_2, your breathing rate would:
   a) increase
   b) decrease
   c) not be affected, as CO_2 is non-toxic

601. CH_3Cl and Freon require oil of:
   a) high viscosity
   b) low viscosity
   c) viscosity doesn't matter

602. Carrene 2 is a:
   a) high-pressure refrigerant
   b) low-pressure refrigerant

603. Which has the highest tensile strength?
   a) copper
   b) aluminum
   c) mild steel

604. Which refrigerant would experience less oil problems?
   a) Freon 12
   b) Freon 22
   c) Freon 113

605. As an oil becomes more diluted with refrigerant, its viscosity will:
   a) increase
   b) decrease
   c) remain the same

606. When an air conditioning system is pumped down and the power is shut off, the suction pressure rises very rapidly. This is an indication that:
   a) the discharge valves are bad
   b) the suction valves are bad
   c) the system is overcharged

607. In ice-making, the flow of the brine is:
   a) through the shell
   b) through the tubes
   c) around the shell

608. Which is more corrosive?
   a) NaCl
   b) Sodium Bromide
   c) CaCl

609. A rise in temperature at the remote bulb of a TXV causes the valve to:
    a) open
    b) close
    c) hunt
610. The common name for Methylene Chloride is:
    a) Carrene 7
    b) Carrene 1
    c) Carrene 2
611. Sub-cooling the liquid before entering the expansion valve:
    a) increases flash gas
    b) decreases flash gas
    c) flash gas is unchanged
612. Which refrigerant is flammable?
    a) CH_3Cl
    b) CO_2
    c) SO_2
613. What type of pump is used to circulate brine to a system?
    a) centrifugal
    b) rotary
    c) reciprocating
614. Viscosity of oil in a Freon system should be:
    a) lower than the temperature of refrigerant used
    b) higher than the temperature of refrigerant used
    c) the same as that used for the refrigerant
615. Frost on the suction line indicates:
    a) liquid carryover
    b) improper installation
    c) it has poorly installed lines
616. A salimeter reads:
    a) relative humidity
    b) dew point temperature
    c) density of the brine
617. C_2Cl and C_2 Bromide are:
    a) extractors
    b) adsorbers
    c) absorbers
618. The clearance between the piston and the head of an NH_3 compressor is:
    a) 1/64"
    b) 1/4"
    c) 1/2"
619. At saturating evaporating pressure, refrigerants for comparison purposes are taken at:
    a) 5°F

b) 0°F
c) 36°F
620. In regard to fire hazards, the safest refrigeration system to use is:
    a) direct expansion with propane
    b) direct expansion with ammonia
    c) indirect system using brine
621. Which would not normally be used on a multiple evaporator system?
    a) low-side float
    b) TXV
    c) capillary tube
622. The expansion valve that responds well to heat loads is:
    a) TXV
    b) AXV
    c) capillary tube
623. The external equalizer line would usually be located:
    a) last coil of the evaporator near the control bulb location
    b) last coil of the evaporator but real close to the compressor
    c) tied into superheat line of the expansion valve for absolute control
624. Which expansion valve keeps an evaporator more flooded?
    a) TXV
    b) AXV
    c) low-side float
625. An accumulator is used to:
    a) discharge vapor from going directly to the condenser
    b) keep the evaporator flooded and separate liquid from vapor
    c) sub-cool the liquid
626. In appearance, an accumulator looks like:
    a) a shell and tube condenser
    b) a vertical shell and tube condenser
    c) an evaporative condenser
627. In a cold evaporator, the viscosity of the oil would:
    a) increase
    b) decrease
    c) remain the same
628. Liquid refrigerant expanded into the suction side would cause:
    a) high head pressure
    b) low head pressure
    c) a decrease in superheat of the suction gas
629. The king valve is located near the receiver to:
    a) isolate the condenser from the receiver
    b) isolate the receiver from the low-side of the system
    c) throttle, so as to control the flash gas in the liquid line

630. Which of the following devices will pass the leak vapors?
    a) a low-side float
    b) a high-side float
    c) thermostatic expansion valves
631. With a high-side float, a rising float ball:
    a) closes the valve
    b) opens the valve
    c) has no effect on the valve
632. In a system with a low-side float, too much refrigerant:
    a) has no effect on the evaporator
    b) causes flooding
    c) starves the evaporator
633. With a TXV, a rise in superheat:
    a) has no effect on refrigerant flow
    b) increases the refrigerant flow
    c) decreases the refrigerant flow
634. With an automatic expansion valve, an increase in pressure:
    a) opens the valve wider
    b) decreases evaporator capacity
    c) closes the valve
635. Melting ice absorbs:
    a) sensible heat
    b) specific heat
    c) latent heat
636. At saturated condensing temperature, refrigerants for comparison purposes are figured at:
    a) 5°F
    b) 0°F
    c) 86°F
637. In ice-making, the brine level should be:
    a) below the water level in the can
    b) above the water level in the can
    c) 18" below the water level in the can
638. The rating of a system at Standard Ton conditions are:
    a) 10°F evaporator at 86°F condenser
    b) 5°F evaporator at 90°F condenser
    c) 86°F condenser at 5°F evaporator
639. Evaporator pipes that are too large:
    a) cause oil to flow through the compressor
    b) help trap oil
    c) make no difference in oil removal
640. An evaporator fed with a TXV at the bottom will:
    a) operate less if flooded

b) operate more if flooded
   c) make no difference
641. When a thermal bulb of TXV loses its charge, the suction and discharge pressures will be:
   a) unchanged
   b) high
   c) low
642. A properly licensed refrigeration operator may:
   a) charge and evacuate a system
   b) charge a system
   c) not charge a system
643. Finned tubes are likely to be used on:
   a) evaporative condensers
   b) double pipe condensers
   c) air-cooled condensers
644. Finned tubes are not likely to be used on an evaporative condenser if:
   a) they may retard heat transfer
   b) cleaning will be more difficult
   c) they can cause pressure drop
645. The purpose of the expansion valve is to:
   a) meter the flow of refrigerant according to the temperature
   b) meter the flow of refrigerant according to the heat load
   c) bypass the evaporative coil if necessary
646. To control capacity on a VSA compressor, use:
   a) synchronous motors
   b) clearance pockets
   c) an inlet vane in the suction line
647. Less condenser is needed for:
   a) NH_3
   b) Freon
   c) R-40
648. A thermostatic process is:
   a) where heat is added or extracted
   b) an adiabatic process
   c) an isothermal process
649. Absolute humidity is:
   a) actual weight of water in a given amount of air
   b) dew point
   c) dry bulb temperature plus the wet bulb temperature
650. A properly installed heat exchanger has:
   a) an increase in pressure
   b) a decrease in pressure
   c) no change in pressure

651. Eliminators are used on a condenser to:
    a) prevent H_2O carry-over
    b) diffuse the air
    c) prevent a drain draft
652. A properly installed heat exchanger:
    a) increases flash gas
    b) decreases flash gas
    c) has no change in flash gas
653. The refrigerant most mixable with oil is:
    a) NH_3
    b) Freon
    c) CO_2
654. A cross head will be found on:
    a) a VSA compressor
    b) an HDA compressor
    c) a horizontal compressor
655. On an NH_3 compressor, the device used to separate the oil is:
    a) an oil still
    b) a separator
    c) an oil chiller
656. Between the compressor and the condenser, the:
    a) superheat is given up first
    b) latent heat is given up first
    c) sensible heat is given up first
657. Which thermometer is considered the most accurate above -38°F?
    a) alcohol
    b) Hg
    c) spirits
658. If an NH_3 compressor at 24 psi was changed to operate at 4 psi, the refrigeration would produce:
    a) the same
    b) twice as much
    c) six times as much
659. The freezing point for R-12 is:
    a) -32°F
    b) -74°F
    c) -252°F
660. A method of recovering oil for an NH_3 system is oil:
    a) separation
    b) still
    c) chiller
661. Frost accumulation on the evaporator coil of an a/c unit is caused by:
    a) high superheat across the coil

b) excess cfm across the coil
c) low cfm across the coil
662. An evaporator condenser has:
   a) finned tubes
   b) studded tubes
   c) bare pipes, prim surface
663. Which is more efficient?
   a) direct system
   b) indirect system
   c) both are the same
664. Brine is circulated by which type of pump?
   a) centrifugal
   b) reciprocating
   c) agitator
665. When liquid passes through an expansion valve, the pressure drops but:
   a) there is little change in total heat
   b) there is no change in total heat
   c) there is a complete change in total heat
666. A stop valve may be installed in the liquid line if:
   a) a receiver is used
   b) a relief valve is used
   c) the installer wants to
667. Stop valves may be used ahead of the relief valve if:
   a) a dual relief valve is used
   b) a receiver is ASME
   c) none can be used
668. Which refrigerant is similar to R-11?
   a) Carrene 2
   b) R-717
   c) R-49
669. What controls the cabinet temperature in most household refrigerators?
   a) a thermostatic expansion valve
   b) a capillary tube
   c) a thermostat in the frozen foods compartment
   d) a thermostat in the perishable foods compartment
670. In a safety head, the machine suction valves are in the:
   a) safety head
   b) valve plate
   c) neither
671. The freezing point for R-22 is:
   a) -256°F
   b) -252°F
   c) -109°F

672. The best temperature to chill drinking water is:
    a) 35°F to 45°F
    b) 45°F to 55°F
    c) 55°F to 65°F
673. The operator shall have the equipment ready for inspection:
    a) annually
    b) bi-annually
    c) at all times
674. Gas masks shall be located:
    a) near the door, inside the machine room
    b) on the wall
    c) in a convenient place outside the machine room
675. In a direct system:
    a) for air conditioning, more refrigerant is in the shell
    b) for industrial, more refrigerant is in the tubes
    c) in both cases it can be either shell or tube
676. A thin layer of carbonate on the inside of the pipe would:
    a) be corrosive
    b) restrict heat transfer
    c) make the pipe last longer
677. Which type of condenser gives the higher efficiency for less floor space?
    a) atmospheric
    b) shell and tube
    c) vertical shell and tube
678. Which would be the most common compressor on the market?
    a) 6 x 6 at 750 rpm
    b) 3 x 3 at 1500 rpm
    c) 6 x 6 at 1500 rpm
679. The code deems it mandatory that all pressure relief devices one-half inch in size or over must be:
    a) approved by any nationally recognized testing laboratory
    b) constructed to meet ASME code requirements
    c) constructed of the same material as the vessel
680. Water will absorb more NH_3 at:
    a) 80°F
    b) 100°F
    c) 200°F
681. A thin coating of $CaCO_3$, spread over the condenser pipes, would:
    a) cause corrosion
    b) prevent corrosion
    c) have no effect on corrosion
682. The heavier the brine:
    a) the higher the specific heat

b) the less the specific heat
c) specific gravity has no effect on the specific heat
683. With a compound compressor, the larger cylinder discharges to the:
   a) high-side of the system
   b) low-side of the system
   c) small cylinder
684. The water regulating valve is connected to the:
   a) low-side of the system
   b) high-side of the system
   c) crankcase of the compressor
685. A high-pressure cut-out is connected to the:
   a) low-side of the system
   b) high-side of the system
   c) crankcase of the compressor
686. To prevent bacteria and mold in vegetables, the temperature should be:
   a) 0°F
   b) 45°F
   c) -15°F
687. The tubes on a horizontal shell and tube condenser are held rigid and made gas tight by:
   a) welding
   b) rolling and flaring
   c) threading
688. A single cylinder acting compressor (horizontal) needs at least:
   a) 2 valves
   b) 4 valves
   c) 8 valves
689. Sub-cooled liquid fed to an expansion valve:
   a) decreases the refrigeration effect
   b) increases the refrigeration effect
   c) does not change the refrigeration effect
690. The head pressure in a CO_2 system would probably be:
   a) 72 atmospheric
   b) 35 atmospheric
   c) 1020 atmospheric
691. The correct temperature for brine for ice-making is:
   a) 14°F to 15°F
   b) 10°F to 15°F
   c) 0°F to 10°F
692. When used to interconnect the high- and low-sides, a heat exchanger's primary function is to:
   a) sub-cool the suction gas
   b) raise the head pressure
   c) pre-cool the liquid refrigerant

693. Which is considered to be in the main electrical circuit of a domestic unit?
   a) plug-in cord and relay
   b) plug-in cord, relay and motor
   c) plug-in cord, thermostat, relay and motor
   d) plug-in cord, thermostat, relay, motor and accessories
694. What is the adjustable pressure range for water-regulating valves for R-12?
   a) 150 to 260 psig
   b) 60 to 160 psig
   c) 0 to 110 psig
   d) 20 to 90 psig
695. What would be the superheat of the suction line of an evaporator if the expansion valve bulb reads 47°F and the evaporator pressure is 26.1 psig, using R-12?
   a) 50°F
   b) 10°F
   c) 12°F
   d) 20°F
696. On a R-22 system, if the condenser water temperature is 75° at the inlet, the compressor discharge pressure should be:
   a) 75 to 100 psig
   b) 125 to 150 psig
   c) 160 to 190 psig
697. Overload heaters should be selected at what percent of full load amperage?
   a) 100%
   b) 115%
   c) 125%
   d) 150%
698. To accurately determine the amount of refrigerant required for a system:
   a) always use 3 pounds per ton of refrigeration
   b) always use 5 pounds per ton of refrigeration
   c) use the volume of receiver
   d) add the pounds required for each component
699. Determine the re-heat coil capacity in heating air from 50°F db, 49°F wb to 62°F db, 54°F wb using 1000 cfm:
   a) 12,960 Btu
   b) 14,521 Btu
   c) 16,819 Btu
   d) 18,314 Btu
700. What is the smallest amount of condenser water at 85°F entering temperature that could be used on a Trane 10 ton unit?
   a) 14.0 gpm
   b) 18.5 gpm
   c) 24.0 gpm
   d) 36.0 gpm

701. If 1/3 of the air supplied to an air conditioning unit is outside air at 95°F db, 75°F wb, and 2/3 is return air at 80°F db, 67°F wb, what is the dry bulb temperature of the mixture?
    a) 83°F
    b) 85°F
    c) 87°F
    d) 90°F
702. If an hp compressor draws 4 kW using 220 volt, 3 phase current, what do the amperes draw per phase?
    a) 10 amps
    b) 12.5 amps
    c) 15 amps
    d) 17.5 amps
703. Shortage of refrigerant is not indicated by which of the following?
    a) high superheat temperature
    b) high liquid line temperature
    c) low suction temperature
    d) solid flow of refrigerant in the sight glass
704. Make-up water is commonly available from the city water supply at:
    a) 30 to 75 ft head
    b) 70 to 125 ft head
    c) 175 to 300 ft head
705. In an average installation, the water tower lowers the condenser water temperature within degrees of local wet bulb temperature:
    a) 5°F to 15°F
    b) 15°F to 25°F
    c) 25°F to 35°F
706. On the average tower, increasing the water flow from 3 gpm to 4 gpm increases the lower capacity:
    a) 5%
    b) 10%
    c) 15%
    d) 20%
707. The conversion factor for changing pressure in psi to feet of water is:
    a) 3.21
    b) 2.13
    c) 3.12
    d) 2.31
708. The freezing point for R-500 is:
    a) -108°F
    b) -256°F
    c) -252°F

709. The formula for Charles' Law is:
    a) $P_o \times T_n = P_n \times T_o$
    b) $P_a \times T_t - T_a \times P_t$
    c) $V_o \times P_n = V_n \times P_a$
    d) both b & c are correct
710. Substances exist in how many states?
    a) one physical form
    b) two physical forms
    c) three physical forms
    d) four physical forms
711. Pressures above atmospheric are measured in:
    a) degrees Celsius
    b) degrees Fahrenheit
    c) psig
    d) inches of mercury
712. Atmospheric pressure is:
    a) 14.7 feet
    b) 14.7 psia
    c) 14.7 psig
    d) 14.7 inches of Hg
713. The symbol Hg indicates:
    a) halogen inches
    b) feet of mercury
    c) hydrogen gallons
    d) inches of mercury
714. Water columns are used to measure:
    a) hose pressure
    b) volume
    c) small pressures, above or below atmospheric
    d) square areas
715. Compound gauges measure:
    a) pressures above and below atmospheric
    b) pressures above atmospheric only
    c) pressures below atmospheric only
    d) density
716. Density is defined as:
    a) pressure exerted on supporting surfaces
    b) the length per unit volume
    c) the weight per unit volume
    d) the same as specific gravity
717. The lowest temperature believed obtainable is:
    a) -400°F
    b) -640°F
    c) -260°F

d) -460°F
718. The "Gas Law Formula" is:
   a) $V_o \times T_n = V_n \times T_o$
   b) $P_n \times T_o = T_n \times P_o$
   c) $P_o \times V_o = P_n \times V_n$
   d) $(P_o \times V_o) \div T_o = (P_n \times V_n) \div T_n$
719. Entropy is:
   a) the heat available measured in Btus per pound per degree change
   b) something used every day by servicemen
   c) a condition of critical pressure and temperature
   d) not used in engineering calculations
720. Boyle's Law, Charles' Law and the Gas Law all deal with:
   a) saturated conditions
   b) relative humidity
   c) sub-cooling
   d) gases only
721. The formula for Boyle's Law is:
   a) $P_o \times T_n = P_n \times T_o$
   b) $V_o \times T_n = V_n \times T_o$
   c) $P_o \times V_o = P_n \times V_n$
   d) none of the above are correct
722. Relative humidity is:
   a) a point where air will not hold any more moisture at 60%
   b) not the same for different temperatures
   c) the percentage of moisture in the air compared to the amount it could hold at the same pressure and temperature
   d) not affected when air is passed over a cooling coil
723. Isothermal expansion and contraction are:
   a) co-efficients of performance
   b) actions which take place without a temperature change
   c) conditions of expansion and contraction of pipe
   d) related to saturated conditions
724. The term "adiabatic" means:
   a) a liquid expands without any heat loss or gain
   b) a term used in cryogenics
   c) the critical temperature had been reached
   d) a gas expands or contracts without any heat loss or gain
725. The valve plate from an open type or semi-hermetic compressor can be removed without disconnecting the compressor from a charge system by:
   a) back-seating both valves
   b) back-seating the discharge valve and front-seating the suction valve
   c) front-seating the discharge valve and back-seating the suction valve
   d) front-seating both valves

726. If the load on a compressor increases, the following will occur:
    a) both suction and discharge pressures will increase
    b) suction will decrease, discharge will increase
    c) suction will increase, discharge will decrease
    d) both will decrease
727. If the outlet valve on a receiver is shut off, the following will occur (assume operating compressor):
    a) the high pressure control will stop the system
    b) the unit will "pump-down"
    c) the suction pressure will increase
    d) the valves and/or gaskets will break on the compress (assume unit has a correctly sized receiver and proper refrigerant charge)
728. When preparing a system to change a cycle component (such as a drier), it is good practice to:
    a) maintain normal operating pressure differences between the high and low sides of the system
    b) pump all the refrigerant into the receiver and shut it off tightly
    c) bleed refrigerant through the system while the drier is being changed at or about 2 psig pressure
    d) let a little air into the system, it won't hurt anything
729. If you front-seat a suction valve on an operating unit:
    a) the compressor is likely to blow up
    b) you can do so if you do it very gradually without hurting the compressor
    c) the discharge pressure will increase to dangerous limits
    d) you should do it rapidly to prevent the oil from boiling
730. Which of the following switches cannot be used in a switching circuit placed in the hot line of a 120 volt single-phase power supply:
    a) SPST
    b) DPST
    c) three switches of an electromagnetic relay
    d) three switches connected in series
731. A control circuit must be low voltage:
    a) true
    b) false
732. What is a pyrometer?
    a) a device used to measure the saturation pressure of the refrigerant vapor
    b) a device used to measure specific weights
    c) a unit used for measuring the specific gravity of a liquid
    d) a thermometer that senses heat, changes it to electrical energy and moves a meter that indicates temperature
733. The centigrade (C) scale is:
    a) presently in common use by serviceman here
    b) a metric system pressure scale

c) a metric system temperature scale
d) cannot be converted to Fahrenheit

734. The gravity pull on an object will give the object a falling acceleration of:
   a) 32.2 feet/second, per second
   b) 32.2 feet/second per 100 feet fall
   c) 32.2 foot/second per foot fall
   d) 32.2 feet/second per minute

735. What is mass?
   a) the most common kind of confusion, such as mass confusion
   b) that which is measured in pounds
   c) the property of all matter, for everything has mass
   d) none of the above is correct

736. If food is frozen slowly at or near the freezing temperatures of water, the ice crystals formed will be:
   a) small in size
   b) there won't be any ice crystals
   c) large in size
   d) none of the above

737. A 34-foot column of water would equal how many inches of mercury?
   a) 50.00" Hg
   b) 29.92" Hg
   c) 2.992" Hg
   d) 299.2" Hg

738. All matter can exist in what states?
   a) all of the following
   b) solid
   c) vapor
   d) liquid

739. What is pressure?
   a) volume per unit area
   b) space used per unit area
   c) force per unit area
   d) shape per unit area

740. What is a solid?
   a) any substance that always takes the shape of its container
   b) any substance which must be contained in a sealed container or it will soon dissipate
   c) both A and B above are correct
   d) a substance which retains its own shape without support

741. A perfect vacuum may be expressed as:
   a) 0 lbs/in² absolute
   b) 0 lbs/in² gauge
   c) 25" Hg
   d) 15" Hg

742. One atmosphere equals:
   a) 14.7 psig
   b) 0 psia
   c) 29.4 psig
   d) 0 psig
743. The force supporting a solid is always:
   a) downward
   b) sideways
   c) upward
   d) backward
744. How many square inches are there in one square foot?
   a) 144 in^2
   b) 576 in^2
   c) 553 in^2
   d) 594 in^2
745. A mercury barometer is used to measure:
   a) pressures in psia below sea level
   b) pressures in gauge absolute
   c) atmospheric pressure
   d) strength of mercury
746. To convert psig to psia, add:
   a) 15 to the psia reading
   b) 15 to the psig reading
   c) 29.92 to the psig reading
   d) 29.92 to the psia reading
747. Atmospheric pressure on one in^2 of surface on the earth at sea level is:
   a) 14.69 psia
   b) 0 psia
   c) 14.7 psi
   d) 14.69 psig
748. The device used for measuring small pressure differences is the:
   a) anemometer
   b) pyrometer
   c) psychrometer
   d) manometer
749. Critical temperature of a substance is:
   a) of no particular interest in refrigerants
   b) the same as critical pressure
   c) the maximum temperature at which a substance may be liquified, regardless of pressure
   d) the point above which all refrigerants operate
750. Temperature work below -250°F is properly called:
   a) cryogenics
   b) areas of high boiling point refrigerants

c) low temperature work
d) hard to come by temperature

751. Dalton's Law of partial pressure expresses:
   a) a condition where two or more gases are present
   b) heat flow in a refrigeration system
   c) a condition where new refrigerants are formed
   d) the law of saturation

752. A hygrometer is a device used to:
   a) measure air volumetric efficiency
   b) meter hygroms
   c) describe a compression system
   d) measure relative humidity

753. Critical pressure of a substance is:
   a) the pressure above which all evaporators should operate
   b) usually an extremely low temperature
   c) at the same point as the critical temperature
   d) the point at or above which a liquid cannot be evaporated, regardless of heat applied

754. Enthalpy is:
   a) the means of finding a pressure
   b) the amount of heat in one pound of a substance
   c) the same as entropy
   d) a base temperature

755. "Pressure drop" is:
   a) the pressure built up in the cylinder at top dead center
   b) the difference in pressure between the inlet and outlet of an object
   c) the pressure in the receiver after the unit has been shut down for a while
   d) the pressure in the compressor cylinder at the bottom of the piston stroke

756. Mechanical cycle is a:
   a) system that operates on the absorption principle
   b) system that never needs a condensing unit
   c) refrigeration serviceman's mode of transportation
   d) cycle which is a repetitive series of mechanical events

757. A relief valve is a:
   a) manually operated valve that controls the amount of heat load the compressor has to take care of
   b) safety valve designed to open before a dangerous pressure is reached
   c) valve used to release air into the system if additional pressure is needed
   d) valve used to reverse the direction of the refrigerant flow depending on whether heating or cooling is desired

758. Gear and diaphragm compressors:
   a) are not commonly used
   b) are a good way of compressing gas
   c) use throw eccentric type crankshafts

d) are used on large-sized jobs
759. The condenser is usually made of:
   a) steel and plastic
   b) copper or steel
   c) rubber and copper
   d) a material similar to glass
760. Condensers can be:
   a) either air-cooled or water-cooled
   b) air-cooled only
   c) water-cooled only
   d) static only
761. What is a sling psychrometer?
   a) a wall-mounted recorder that has a built-in sling to catch particles heavier than air (chromets)
   b) a humidity-measuring device with wet and dry bulb thermometers which is moved rapidly through the air when measuring humidity
   c) a strap-type sling for mounting compressors on the wall
   d) none of the above
762. A cooling tower is:
   a) a platform the compressor is placed on to keep it cool while operating
   b) a tower the condenser is placed in to allow better heat transfer for the removal of latent heat of condensation
   c) a device which cools water by water evaporation in air
   d) a device the compressor oil is pumped through to cool it down to a better operating temperature
763. What is latent heat?
   a) the heat required to raise the temperature of a vapor
   b) the heat given off in the compressor as heat of compression
   c) the heat required to change the state of a substance
   d) the heat that you measure with a thermometer
764. Where is the suction line on a system?
   a) between the compressor and condenser
   b) between the evaporator and compressor
   c) between the condenser and metering device
   d) none of the above
765. What is psia?
   a) gauge pressure minus atmospheric pressure
   b) atmospheric pressure minus gauge pressure
   c) gauge pressure divided by atmospheric pressure
   d) gauge pressure plus atmospheric pressure
766. What does lapping mean?
   a) to smooth a metal surface to a high degree of refinement or accuracy using a fine abrasive

b) the proper way of lifting a heavy object off the floor by first placing it in your lap before standing upright
c) the process of re-packing the service valves
d) to straighten out the shank on bent reed valves

767. What is the valve plate?
    a) the part of the compressor to which the suction valve is connected
    b) the part of the compressor located between the top of the compressor body and head that contains the compressor valves
    c) part of the compressor to which the discharge service valve is connected
    d) the section of the compressor that contains the valves for the oil pump

768. What is a ton of refrigeration?
    a) a refrigeration system that weighs 2,000 pounds
    b) the refrigeration effect equal to the melting of one ton of ice in 12 hours
    c) the refrigeration effect equal to the removal of 288,000 Btu in 12 hours
    d) the refrigeration effect equal to the melting of one ton of ice in 24 hours

769. What is refrigerant charge?
    a) the electrical charge placed in a refrigerant to increase its heat-absorbing ability
    b) the amount of refrigerant needed in a system
    c) the wholesaler's cost of the refrigerant
    d) how much heat a refrigerant will absorb or give off to change its state

770. What is the purpose of the liquid receiver?
    a) to change the high pressure liquid to a low pressure liquid
    b) to control the amount of refrigerant flowing into the condenser
    c) to store excess liquid refrigerant
    d) to increase the volumetric efficiency of the compressor

771. What is a high-side float?
    a) metering device used on a critically charged system that controls the level of the liquid refrigerant in the high pressure side of the system
    b) a float type of switch that stops the compressor when the pressure on the high-side becomes too high
    c) a float type of valve that controls the amount of refrigerant stored in the receiver
    d) a float type of valve that keeps the oil in the crankcase level

772. What is a foot-pound?
    a) a foot that weighs 16 ounces
    b) a pound that is a foot long
    c) one pound moved a distance of one yard (3 feet)
    d) one pound moved a distance of one foot

773. What is hot gas by-pass?
    a) the piping system which moves the refrigerant gas from the condenser into the low pressure side of the system

b) a piping system that allows hot gas to flow directly into the conditioned space if it should get too cold
c) a piping system that directs hot gas directly to the condenser from the compressor
d) a section of the hermetic-type compressor used to help cool the motor windings

774. What is used to flame test for leaks?
a) halide leak detector
b) an electronic leak detector
c) an ammonia swab
d) a soap and water solution

775. What is an oil separator?
a) a device that separates the oil from the moisture in a system
b) a device that settles the wax from the oil
c) none of these answers is correct
d) a device used to separate the oil from the gaseous refrigerant

776. What is a stationary blade compressor?
a) a compressor that has a stationary fan blade
b) a compressor that has the blades inside the rotor and rotates with it
c) a reciprocating compressor mounted securely to its base
d) a rotary pump that uses a stationary blade and rides against the rotor

777. What is the Rankin Scale?
a) the name given for the absolute centigrade scale
b) a scale used for measuring the amount of acid in water
c) a scale used for measuring the amount of acid inrefrigerant
d) the absolute Fahrenheit scale where zero is actually -460°F

778. What is the Kelvin Scale?
a) the name given to the absolute Fahrenheit scale
b) a scale for measuring metric weights
c) the name given to the absolute centigrade scale
d) none of the above

779. Where is the liquid line located in the system?
a) between the compressor and condenser
b) between the evaporator and compressor
c) none of these answers is correct
d) between the condenser and the metering device

780. Latent heat of fusion is:
a) the heat released in changing a substance from a liquid state to a solid state
b) the heat released in changing a substance from a vapor state to a liquid
c) the heat absorbed in changing a substance from a liquid state to a vapor
d) the heat absorbed in the liquid refrigerant between the receiver and metering device

781. The compressor's job is to:
a) pump the liquid refrigerant around

b) move gas from the low temperature to the high temperature side of the system
c) pump heat into the evaporator
d) relieve the discharge line of excessive pressure

782. The condenser:
    a) is located between the compressor and the metering device
    b) is located between the liquid line and the evaporator
    c) takes heat into the system
    d) reduces pressure

783. A receiver is a device:
    a) located between the liquid line and the compressor
    b) not needed on all refrigeration systems
    c) for getting rid of all heat absorbed into the system
    d) for storing superheated gas

784. The evaporator:
    a) is located between the compressor and receiver
    b) is located between the compressor and condenser
    c) puts the heat into the refrigerant
    d) gets rid of heat

785. The refrigerant:
    a) is not needed in a mechanical refrigeration system
    b) is transferred from the inside to the outside of the system
    c) gets rid of heat by radiation
    d) transfers heat from one side of the system to the other

786. The discharge line:
    a) connects the metering device to the evaporator
    b) discharges heat to the metering device
    c) discharges the gas from the evaporator to the condenser
    d) is located between the compressor and condenser

787. Heat enters the evaporator through:
    a) radiation only
    b) radiation, convection and conduction
    c) heat generated by the compressor
    d) the insulation in the outside walls only

788. Flash gas is a condition of the:
    a) metering device
    b) compressor
    c) condenser
    d) receiver

789. The suction line connects:
    a) the condenser and the receiver
    b) two parts of the system and is the smallest size of line used
    c) the evaporator to the compressor
    d) the metering device to the evaporator

790. The largest size line used on a compression system is:
    a) the discharge line
    b) the suction line
    c) lines are all the same size
    d) the liquid line
791. A gas is superheated as it leaves the:
    a) metering device
    b) discharge line
    c) condenser
    d) evaporator
792. The pressure in the liquid line is:
    a) the same as crankcase pressure
    b) the same as the suction line
    c) the same as the evaporator
    d) usually the same as condenser pressure
793. The condition of the gas leaving the compressor is:
    a) sub-cooled
    b) superheated
    c) saturated
    d) flash gassed
794. A saturated condition exists only in the:
    a) evaporator and condenser
    b) suction line and compressor
    c) receiver and liquid line
    d) end of the evaporator and the suction line
795. As liquid leaves the metering device to the evaporator:
    a) the pressure is reduced and the liquid is superheated
    b) it is stored in the receiver
    c) part of the liquid is vaporized, thus cooling the rest of it
    d) expansion occurs in the condenser
796. Heat flow in a mechanical system is:
    a) into receiver and out of evaporator
    b) into evaporator and out at condenser
    c) into liquid line and out at receiver
    d) into compressor and out at the liquid line
797. The compressor adds heat to the system by:
    a) reducing the volume of the gas and raising its pressure
    b) this is not known as heat of compression
    c) taking the heat of the condenser and compressing it
    d) raising its temperature by specific heat
798. The most heat flows out of a refrigeration system through the:
    a) line connecting condenser to receiver
    b) suction line

c) discharge line
d) liquid line

799. The condition of the fluid leaving the condenser:
    a) is in a gaseous state
    b) is in a solid state
    c) is slightly superheated
    d) should be slightly sub-cooled

800. The condenser and discharge line:
    a) need help in getting rid of heat
    b) connect the evaporator and suction line
    c) must rid the system of all heat picked up in the cycle
    d) take heat into the system

801. The line connecting the receiver to the evaporator is:
    a) the suction line
    b) the liquid line
    c) the discharge line

802. The discharge service valve is located:
    a) usually in a valve plate
    b) on the inside of the compressor
    c) on the outside of the compressor and is a service valve
    d) in the piston

803. A receiver is used to:
    a) store liquid refrigerant
    b) supply liquid to the compressor
    c) connect the line to the evaporator
    d) measure the Btu absorbed in the system

804. What color of the halide torch flame indicates a leak?
    a) red
    b) blue
    c) yellow
    d) green
    e) colorless

805. How many basic types of relays are used on domestic units?
    a) one
    b) two
    c) three
    d) four
    e) five

806. Why must one be careful of the amount of refrigerant put into a cylinder?
    a) always allow for the expansion of the gas
    b) always allow for the expansion of the cylinder
    c) always allow for the expansion of possible moisture
    d) the cylinder will burst if full of liquid
    e) refrigerant is expensive

807. Does the service valve attachment have threads on it?
    a) yes
    b) no
    c) on some models
    d) only if it is a high-pressure unit
    e) no valve system is used

808. Of what materials are most service valve stems made?
    a) steel
    b) copper
    c) brass
    d) monel
    e) tool steel

809. What type of relay does not need electromagnets?
    a) magnetic type
    b) voltage type
    c) amperage type
    d) hot wire type
    e) weight type

810. What may be wrong if the relay takes too long before the starting circuit is opened?
    a) the capacitor is shorted
    b) the starting winding is grounded
    c) the voltage is too high
    d) the unit is overloaded
    e) the unit is too cold

811. How does an electronic leak detector indicate a refrigerant leak?
    a) a flame color change
    b) a meter reading, light or bell
    c) a color change in the tube
    d) a color trace
    e) bubbles

812. What service operations are best performed with the aid of a high-vacuum pump?
    a) setting a motor control
    b) checking the amount of refrigerant in a system
    c) dehydration of a system
    d) setting the thermostatic expansion valve

813. What is probably wrong if the cabinet is too cold and the unit runs continuously?
    a) faulty thermostat
    b) faulty relay
    c) nothing
    d) the unit is out of refrigerant
    e) an open circuit

814. What is the most common indication of a shortage of refrigerant in a hermetic system?
    a) excessive head pressure
    b) shorter running part of cycle
    c) lowering of frost line on the evaporator
    d) warmer cabinet
    e) excessive frosting
815. How may oil be added if the system cannot be made to produce a vacuum?
    a) it cannot
    b) use a hand pump
    c) build up pressure in an oil-charged cylinder
    d) remove all the refrigerant
    e) it is not important to have the correct oil charge
816. What is the most important thing to do if a shortage of refrigerant is discovered?
    a) charge the unit
    b) stop the unit
    c) overhaul the unit
    d) find the leak
    e) check the quality of oil in the unit
817. What precautions must be taken when adding oil to a system which has service valves, and the vacuum method (using a bottle) is used?
    a) always keep the end of the hose submerged in oil
    b) draw a vacuum, open the unit and pour the oil in through a funnel
    c) produce a vacuum and add oil to the high-side of a system
    d) remove all refrigerant, then add oil
    e) always purge the high-side of the system of air first
818. What is the purpose of the relay?
    a) it permits electricity to flow through the starting winding until the motor reaches 2/3 of its speed and then opens the starting winding
    b) it permits electricity to flow through the running winding until the motor reaches 2/3 of its speed and then opens the running winding
    c) to prevent the overheating of the running winding
    d) to prevent burnout from high temperatures
    e) to control the box temperature
819. What will liquid refrigerant do to the eyes and skin?
    a) burn the eyes and skin
    b) freeze the eyes and skin
    c) irritate the eyes and skin
    d) nothing
    e) lubricate the eyes
820. What permits the service valve attachment to be swiveled to any position?
    a) tapered threads
    b) a swivel nut on the valve attachment

c) the attachment fitting has tapered threads
d) it cannot be swiveled
e) a compressible gasket

821. What kind of threads are used in the gauge opening of a service valve?
    a) pipe
    b) NF
    c) NC
    d) national extra fine
    e) any kind of thread

822. How does the piercing needle provide a leak-proof joint when turned all the way in?
    a) it does not
    b) a synthetic rubber gasket
    c) the needle seats on the walls of the pierced hole
    d) a cap is used
    e) a 45° flare

823. A "frozen" valve stem be loosened by:
    a) using a larger wrench
    b) hammering on it
    c) heating the valve body
    d) using a pipe wrench
    e) using oil

824. What circuit is energized after the motor reaches operation speed?
    a) both run and start windings
    b) only the run winding
    c) only the start winding
    d) none of them
    e) only the relay circuit

825. How does the leaking refrigerant reach the halide flame?
    a) it is mixed with the fuel
    b) it is drawn through a sniffer tube
    c) a pump is used
    d) an aspirator is used
    e) the refrigerant is in the air around the flame shield

826. What instrument is often combined with the high vacuum pump as an assembly?
    a) a high-vacuum gauge
    b) a drier
    c) a gauge manifold
    d) a filter
    e) an acid indicator

827. How does one check the electrical power outlet for the unit?
    a) ammeter
    b) test light

c) if cabinet light won't work, power is off
d) if house lights work, the power is all right
e) ohmmeter

828. What is a good vacuum, measured in millimeters of mercury column?
   a) 5 to 10
   b) 25 to 35
   c) 40 to 50
   d) 55 to 65
   e) 70 to 80

829. Why must the condenser surface be kept clean?
   a) to prevent overheating the fan motor
   b) to keep the head pressure down
   c) to prevent dust in the room
   d) for sanitation purposes
   e) to keep moisture out of the system

830. What is considered the best way to heat a service cylinder?
   a) a welding torch
   b) a gas-air torch
   c) a blow torch
   d) hot water
   e) hot gas

831. What is the primary purpose of oil in a refrigeration system?
   a) cooling
   b) lubrication
   c) removal of heat
   d) to act as a catalyst to the refrigerant
   e) to prevent oxidation of parts

832. What is the basis of operation of the potential relay?
   a) an increase in voltage as the unit approaches and reaches its rated speed
   b) a decrease in amperage as the unit approaches and reaches its rated speed
   c) the difference in magnetic impulses
   d) the potential difference between the two magnetic forces e) the operation of a magnetic coil connected to the solenoid

833. Why should the low-side pressure be kept within reasonable limits when charging refrigerant into the low-side?
   a) to keep from overworking the compressor
   b) to prevent liquid pumping
   c) to speed up the operation
   d) to prevent freezing the cylinder
   e) to prevent rupturing the cylinder

834. What is one of the chief properties required of a good refrigerant oil?
   a) it must separate from the refrigerant at high temperatures
   b) it must circulate with the refrigerant
   c) it should separate at low temperatures

d) it must have high viscosity
e) it must have high pour point

835. In what position are the points in a potential relay during the off cycle?
    a) closed
    b) midway
    c) open
    d) this is dependent upon the unit
    e) there are no points

836. Why does heating the cylinder with hot water speed up the charging operation?
    a) warm gas pumps better
    b) the liquid evaporates faster
    c) it removes the oil from the refrigerant
    d) it keeps the charging line clean
    e) it warms the compressor

837. How is maximum refrigerant flow obtained through a Schrader or dill valve core?
    a) core all the way out
    b) core depressed half way
    c) core removed
    d) core all the way in
    e) core depressed even with fitting

838. When should the packing nut be loosened?
    a) only if the valve stem turns with difficulty
    b) whenever the valve is used
    c) only to replace the packing
    d) when the packing leaks
    e) when the valve stem breaks

839. What special precautions must be taken with a weight-type operated relay?
    a) it must be carefully leveled
    b) it must be kept cool
    c) it must be over capacity
    d) it must be noise insulated

840. What happens when a two-way service valve stem is turned all the way out?
    a) it will come out
    b) it closes the gauge opening
    c) it closes the refrigerant line
    d) it is front-seated
    e) nothing

841. Why must relay covers be kept as tight as possible?
    a) it is not necessary
    b) dust will cause the points to deteriorate
    c) to keep the unit cool

d) to keep the unit warm
   e) to minimize relay noise
842. Why must the alcohol fuel be very clean?
   a) to obtain a good flame color
   b) to enable the fuel to go through the small passage
   c) to prevent corrosion
   d) to build up enough pressure to keep the flame burning
   e) to keep the alcohol from exploding
843. When installing a water regulating valve on a refrigeration system, it should be located:
   a) on the water inlet side
   b) on the water outlet side
   c) either at inlet or outlet
844. Under most conditions, the most preferable method of capacity control for a centrifugal compressor is:
   a) suction line throttling dampers
   b) condenser temperature control
   c) variable speed drive
845. A refrigerant that is flammable is:
   a) methyl chloride
   b) sulphur dioxide
   c) carbon dioxide
846. Condensing mediums must be:
   a) in the right temperature range
   b) a liquid
   c) a gas
847. In order to place a system on a pump down cycle, which of the following devices is necessary?
   a) a solenoid valve
   b) a thermostat
   c) a pressure control
848. Some manufacturers use mufflers to:
   a) prevent pulsation
   b) retard noise
   c) sub-cool liquid refrigerant
849. Considering the areas to be all equal, which of the following would permit the greatest heat gain?
   a) wood
   b) glass
   c) brick
850. An awning over a window would be less beneficial on which wall?
   a) south
   b) east
   c) north

851. An oil safety switch operates on:
   a) differential between oil pump discharge pressure and the compressor's suction pressure
   b) differential between oil pump suction pressure and the compressor's discharge pressure
   c) discharge pressure of the oil pump
852. All oil safety switches should be of the type that are:
   a) automatically reset
   b) manually reset
   c) internally reset
853. What size systems, using Freons, have to be vented to the outside atmosphere?
   a) any system over 15 tons
   b) a system containing over 100 lbs of refrigerant
   c) any system containing over 31 lbs of refrigerant
854. A pressure vessel over 6 inches in diameter must be constructed:
   a) by the manufacturer of the refrigerant equipment
   b) according to the underwriters laboratory standards
   c) according to the ASME standards
855. A fusible plug discharges the refrigerant in a system:
   a) if the pressure rises more than the pressure stamping on the fusible plug
   b) if the temperature rises more than the temperature stamping on the fusible plug
   c) in case of fire
856. An increase in compression ratio would:
   a) decrease the efficiency of the system
   b) have no effect on the efficiency of the system
   c) increase the efficiency of the system
857. In which part of the system is oil more soluble?
   a) liquid line
   b) suction line
   c) discharge
858. What is the lowest temperature at which R-12 will boil?
   a) -21°F
   b) below -100°F
   c) 0°F
859. An evaporator of a given size can extract more heat at:
   a) 20° evaporator temperature
   b) 0° evaporator temperature
   c) 40° evaporator temperature
860. A compressor would be required to pump a larger volume of vapor with an evaporator temperature of:
   a) 0°F
   b) 40°F
   c) 20°F

861. System A has a 5° evaporator temperature and an 86° condenser temperature; system B has a 40° evaporator temperature and an 86° condenser temperature. To produce one ton of refrigeration, the compressor needs to handle:
   a) the same weight of refrigerant in both systems
   b) more refrigerant in system A
   c) more refrigerant in system B

862. A water-cooled condensing unit having a rise of 15° in the condensing water requires how many gallons of water per minute for each ton of refrigeration produced?
   a) 1 gallon
   b) 2 gallons
   c) 3 gallons

863. System A has a 5° evaporator temperature and an 86° condenser temperature; system B has a 40° evaporator temperature and an 86° condenser temperature. The compression ratio would be:
   a) the same in both systems
   b) greater in system A
   c) greater in system B

864. Dual pressure relief devices are required when:
   a) two compressors discharge into one condenser
   b) Group II or Group III refrigerants are used
   c) the condenser exceeds 10 cubic feet

865. In a system with an evaporator temperature below zero, it is good practice to install:
   a) an oil separator
   b) an oil trap
   c) a crankcase heater

866. Compressor heaters are installed by the manufacturer to prevent:
   a) frosting of the compressor
   b) the oil from congealing
   c) refrigerant from condensing in the compressor

867. The principal time when a compressor heater should be in operation is:
   a) during the off cycle
   b) during the running cycle
   c) when adding refrigerant

868. A condenser that uses both water and air is called:
   a) a dual purpose condenser
   b) an evaporative condenser
   c) a cooling tower

869. A part of the heat that is removed by the condenser is called:
   a) heat of evaporation
   b) latent heat
   c) heat of condensation

870. Oil traps are installed in the suction line when the condensing unit is:
    a) air-cooled
    b) installed below the evaporator
    c) installed above the evaporator
871. Assuming the bodies of a 10 ton and a 5 ton expansion valve are of the same size, the power element would be:
    a) interchangeable
    b) larger on the 5 ton expansion valve
    c) larger on the 10 ton expansion valve
872. Increasing the superheat setting of a TX expansion valve would:
    a) increase the frost line
    b) not affect the frost line
    c) decrease the frost line
873. Under what condition may bushings be used in fittings?
    a) when the reduction is two or more pipe sizes
    b) when the reduction is one or more pipe sizes
    c) on any fitting
874. ASME stamping is mandatory when a pressure vessel exceeds:
    a) 6" inside diameter
    b) 12" inside diameter
    c) 10 ft^3
875. Crankcase heaters are sometimes used to:
    a) prevent the compressor from freezing
    b) keep the oil viscosity constant
    c) prevent refrigerant from condensing in the compressor
876. Hot gas mufflers are sometimes used on large systems for the purpose of:
    a) removing pulsations
    b) de-superheating the discharge gas
    c) preventing oil leaving the compressor
877. When condenser water temperature increases, the compression ratio would:
    a) decrease
    b) increase
    c) remain the same
878. What type of compressor uses a damper as a means of capacity control?
    a) rotary
    b) centrifugal
    c) reciprocating
879. Under which of the following temperatures would one pound of saturated air contain the greatest amount of moisture?
    a) 25°F
    b) 60°F
    c) 80°F

880. Under which of the following conditions would a compressor be more efficient?
   a) 40 lbs suction psig
   b) 25 lbs suction psig
   c) 150 lbs discharge psig
   d) 125 lbs discharge psig
881. Assuming all conditions to be equal, a given compressor removes more heat when the refrigerant used is:
   a) R-12
   b) R-22
   c) ammonia
882. Liquid injection into the suction line is sometimes necessary in order to cool the compressor:
   a) under full load
   b) under unloaded conditions
   c) upon condenser water failure
883. When installing a piece of air-cooled refrigeration equipment on the roof, a refrigeration permit:
   a) is required
   b) and plumbing permit are required
   c) and building permit are required
884. A relief valve on an ammonia system must be discharged:
   a) into the room
   b) to the atmosphere
   c) above the roof level
885. The type of valve that has the least resistance to flow is a:
   a) gate valve
   b) globe valve
   c) shut-off valve
886. A properly installed refrigeration line should be:
   a) installed level
   b) pitched toward the compressor
   c) pitched in the direction of flow
887. When installing refrigerant lines, it is recommended that oil traps be used:
   a) on all vertical lines
   b) when the suction lines rise 8 feet or more
   c) when the condenser is on the roof
888. The greatest permissible pressure drop allowed would be in the:
   a) discharge line
   b) suction line
   c) liquid line
889. The approximate pressure drop in the vertical rise of a liquid line of a freon system is:
   a) 0.50 lbs/ft

b) 0.75 lbs/ft
   c) 1.50 lbs/ft
890. What type of control is usually used for loading or unloading a compressor automatically?
   a) pressure switch
   b) thermostat
   c) manual valve
891. When an oil separator is used on a Freon system, the oil is:
   a) returned to the compressor crankcase
   b) drained into a receptacle
   c) returned with the suction vapor
892. The minimum velocity of refrigerant in a hot gas line should be:
   a) 1500 fpm
   b) 500 fpm
   c) 1200 fpm
893. When a refrigerant line rises vertically, the minimum velocity, compared to a horizontal line, should be:
   a) increased
   b) the same
   c) decreased
894. A water regulating valve shall be installed:
   a) upstream from the first component installed
   b) either at the inlet or the outlet of the condenser
   c) in the top of the condenser
895. The heat gain due to solar heat is greatest on the south wall of a building:
   a) at the equator
   b) at 30° latitude
   c) at 40° latitude
896. A building painted which of the following colors would have the greatest cooling requirements?
   a) dark green
   b) white
   c) red
897. For each person seated at rest, how many Btu per hour are added to the heat load?
   a) 400
   b) 650
   c) 1,000
898. Which of the following relative humidity conditions would have the greatest heat load?
   a) 10%
   b) 50%
   c) 90%

899. Dew point is:
   a) the point at which air begins to give up its moisture
   b) 32°F
   c) the same as the wet bulb temperature
900. Relative humidity (rh) is:
   a) the comfort zone
   b) grains of moisture in the air
   c) the amount of moisture in the air compared to 100% saturated air
901. Under which of the following conditions is a pressure relief device required on the low-side of a system?
   a) any time the evaporator is located downstream from the heat-producing equipment in an institutional or public assembly occupancy
   b) any time an accumulator is installed in conjunction with an evaporator
   c) when the system serves an evaporator directly above the machinery room
902. In an institutional occupancy, refrigerant piping may pass between floors of a building under which of the following conditions?
   a) when all concealed pipe joints have inspection openings
   b) for the purpose of connecting to a condenser on the roof
   c) when the system serves an evaporator directly above the machinery room
903. In a commercial occupancy, an installation permit would not be required on:
   a) a new self-contained system of 1 hp or less, 6 lbs or less of refrigeration
   b) any system 1 hp or less
   c) any self-contained system
904. A start kit will:
   a) equal pressure during off cycle
   b) increase resistance in the common leg
   c) increase starting torque
   d) decrease starting torque
905. A start kit for residential equipment contains a:
   a) start capacitor and a run capacitor
   b) step-up transformer and a run capacitor
   c) start capacitor and a potential relay
906. When the unit locks out on the high pressure cut-out, this is a good indication of a restriction in the refrigerant circuit.
   a) true
   b) false
907. An air conditioning system can accomplish more sensible cooling when the air in the conditioned space has a low moisture content.
   a) true
   b) false
908. The primary purpose of a crankcase heater is to:
   a) step 240V AC down to 24V AC
   b) keep refrigerant from condensing in the compressor

c) mechanically add starting torque to a compressor
d) locate oil trapped in a system

909. If you have compressor burn, you should take an oil sample for acid testing.
    a) true
    b) false

910. Insulating the suction line is to prevent:
    a) excessive superheat
    b) excessive sub-cooling
    c) condensate damage
    d) damage to the copper line

911. The thermostatic expansion valve is designed to maintain constant:
    a) flow
    b) temperature
    c) pressure
    d) superheat

912. The sensing bulb on the thermo-expansion valve is attached to the suction line.
    a) true
    b) false

913. A dirty or partially blocked condenser coil will cause the head pressure to:
    a) decrease
    b) remain the same
    c) increase
    d) it would have no effect

914. Ductwork located in unconditioned areas should be insulated.
    a) true
    b) false

915. If a system is properly charged, low outside ambient temperatures do not affect system operation.
    a) true
    b) false

916. When a warm system is first started after a long idle period, it operates under high load conditions until the system pulls down to normal operating temperatures and pressures.
    a) true
    b) false

917. With an increased heat load on the evaporator coil, the thermo-expansion valve will:
    a) completely close
    b) blow a fuse
    c) allow more refrigerant to flow
    d) de-energize the compressor

918. With a thermo-expansion valve, the superheat should be:
    a)   0 to 5
    b)   5 to 15

c) 15 to 25
d) 25 to 35

919. With a capillary tube system, superheat will vary over a wide range depending on the outside temperature.
    a) true
    b) false

920. Where do air and non-condensable gases tend to collect in a system?
    a) condenser
    b) evaporator
    c) compressor
    d) connecting tube

921. Low suction pressure can be caused by:
    a) restriction in the system
    b) a faulty thermo-expansion valve
    c) low air flow through the evaporator
    d) all of the above

922. When an internal mechanical failure occurs, the compressor always makes abnormal noises.
    a) true
    b) false

923. A stuck compressor can be caused by:
    a) a loss of oil
    b) oil trap in system low-side
    c) internal overheating
    d) all of the above

924. On a 24V circuit, when the contacts are shorted to ground, which side of the step-down transformer fails?
    a) primary
    b) secondary

925. A tight running compressor will cause the running amps to be:
    a) higher than normal
    b) lower than normal
    c) this has no effect
    d) amp change depends upon ambient temperature

926. All three-phase equipment must have start capacitors.
    a) true
    b) false

927. When the contactor buzzes, this indicates that no current is flowing to the contactor coil.
    a) true
    b) false

928. Blackened mercury bulbs on a thermostat indicate:
    a) old age of thermostat
    b) high amperage draw exists

c) line voltage has been applied
d) short cycling

929. To check out the blower relay at the thermostat, you must jump the:
   a) R & W terminals
   b) W & G terminals
   c) Y & R terminals
   d) R & G terminals

930. The cooling anticipator produces its "false heat" when the thermostat calls for cooling.
   a) true
   b) false

931. When tests indicate questionable starting relay performance, it is good policy to replace both the relay and the starting capacitor.
   a) true
   b) false

932. The three terminals on a single-phase compressor are:
   a) main, auxilary and ground
   b) primary, secondary and intermediate
   c) start, run and common
   d) normally open, normally closed and common

933. Improper capacitor circuit wiring can cause a compressor motor burn-out.
   a) true
   b) false

934. An air conditioning unit that utilizes a cap tube would come from the factory with what type of compressor motor?
   a) shaded pole
   b) permanent split capacitor
   c) capacitor start
   d) capacitor start-capacitor run

935. The internal overload of the compressor senses:
   a) low voltage at the compressor
   b) amperage draw of the control circuit
   c) excessive fan motor amp draw
   d) excessive amperage in the compressor motor circuit

936. The compressor does not run. You find the condenser fan operating. Reading out the compressor, you get these results: R to S is 5 ohms, R to C is open and S to C is open. You conclude:
   a) compressor contactor faulty
   b) open run winding
   c) internal overload open
   d) open start winding

937. High pressure safety switches will generally close their contacts at:
   a) 300 to 400 psi
   b) 400 to 500 psi

c) 500 to 600 psi
d) none of the above

938. In troubleshooting the low voltage control circuit, we can safely assume the transformer is in good condition if the voltage across the secondary winding is 24 volts.
   a) true
   b) false

939. Compressor motor winding resistances are generally in the range of:
   a) 0 to 10 ohms
   b) 100 to 200 ohms
   c) 1000 to 2000 ohms
   d) all of the above

940. A standard VOM can be used to determine if a capacitor has shorted internally or shorted to ground.
   a) true
   b) false

941. The run capacitor will be electrically connected to the compressor at terminals:
   a) R & C
   b) R & S
   c) S & C
   d) 2 & 5 on potential relay

942. Why is one terminal of a running capacitor identified by a mark?
   a) this is the negative terminal
   b) this is the positive terminal
   c) this terminal usually shorts to the metal case if the capacitor fails
   d) this is the load side terminal

943. When capacitors are wired in parallel, the equivalent microfarad's capacitance value:
   a) increases
   b) decreases
   c) remains constant
   d) differs with start and run capacitors

944. The purpose of a time delay fuse is to avoid nuisance blowing during starting, when the in-rush is much higher than the normal running current.
   a) true
   b) false

945. A low voltage reading across an energized contactor from L to T side indicates the following:
   a) contactor open
   b) excessive resistance
   c) everything Ok
   d) none of the above

946. When the contactor closes but the motor won't start and doesn't hum, the problem is usually in the low voltage circuit.
   a) true
   b) false
947. Air filters should be cleaned or changed:
   a) when necessary
   b) every 6 months
   c) every month
   d) annually
948. A pressure relief device is required on a shell and tube evaporator when:
   a) the refrigerant is in the shell
   b) the refrigerant is in the tubes
   c) the refrigerant is in either the shell or the tubes
949. Specific gravity is:
   a) measured in lbs/in^2
   b) measured in gallons
   c) the same as density
   d) the ratio of weight of an equal volume of water
950. The current-type relay has a switch that is normally:
   a) open
   b) closed
951. The coil on a potential relay is placed between what terminals on the compressor?
   a) R and C
   b) C and S
   c) S and R
952. If the relay terminals are marked 5, 2, 1, it usually is what type of relay:
   a) current
   b) potential
953. A relay can be used with one or more capacitors.
   a) true
   b) false
954. A 2 hp compressor would use what type of relay?
   a) current
   b) potential
955. The water valve on a packaged air conditioner controls:
   a) head pressure
   b) suction pressure
   c) discharge temperature
956. Comparing tower cfm with evaporative condenser cfm, the tower cfm is:
   a) less than an evaporative condenser
   b) equal to an evaporative condenser
   c) greater than an evaporative condenser

957. The condenser's job is to:
    a) supply the cycle with heated gas
    b) feed the evaporator gas
    c) condense refrigerant
    d) feed the compressor low-pressure gas
958. The compressor suction and discharge valves are located:
    a) the same as service valves
    b) can be located in a safety head
    c) between the king and queen valves
    d) usually in the valve plate in the compressor
959. The cooling coil is called a:
    a) condenser
    b) evaporator
    c) freezer
960. The purpose of the evaporator is:
    a) to remove heat
    b) to remove heat of compression
    c) the same as a receiver
961. What does a humidistat control?
    a) moisture
    b) flash gas
    c) refrigeration
962. What is a drier used for?
    a) drying oil
    b) holding flash gas
    c) drying refrigerant
963. Five tons of refrigeration is equal to how many Btu?
    a) 120,000
    b) 90,000
    c) 60,000
964. The freezing point for R-113 is:
    a) -109°F
    b) -31°F
    c) 1252°F
965. The recommended pressure drop for designing piping for towers is how much psi per 100 feet of pipe length?
    a) 1 to 2
    b) 2 to 3
    c) 3 to 4
    d) 3 to 8
966. Make-up water for a cooling tower is supplied by the city usually at a pressure equal to:
    a) 30 to 75 ft
    b) 70 to 175 ft

c) 175 to 300 ft
967. The freezing point for R-764 is:
   a) -109°F
   b) -31°F
   c) -103°F
968. The freezing point for R-11 is:
   a) 74°F
   b) -74°F
   c) -168°F
969. The operating voltage of most defrost timers would be:
   a) 24 volts
   b) 230 volts
   c) 230 volts, three phase
970. The pH value of water in a tower should be kept between the limits of:
   a) 3.5 to 4.5
   b) 4.5 to 5.5
   c) 5.5 to 6.5
   d) 6.5 to 7.5
971. Nominal tonnage ratings for evaporative condensers used for air conditioning units are based on what condensing temperature?
   a) 95°F
   b) 100°F
   c) 105°F
   d) 110°F
972. When checking resistance (in ohms) to identify R-S-C, we know that what will be the smallest?
   a) resistance from R to C
   b) resistance from S to C
   c) resistance from R to S
973. Nominal tonnage ratings for evaporative condensers used for air conditioning units are based on a suction temperature of:
   a) 35°F
   b) 40°F
   c) 45°F
   d) 50°F
974. Nominal tonnage ratings for evaporative condensers used for air conditioning units are based on an entering wet bulb temperature of:
   a) 75°F
   b) 76°F
   c) 77°F
   d) 78°F
975. What type of motor uses a potential relay and a start capacitor (one phase)?
   a) PSC
   b) CSIR

c) CSR
d) SP
976. A mechanical shaft seal is necessary in a:
   a) semi-hermetic compressor
   b) hermetic reciprocating compressor
   c) hermetic rotary compressor
   d) open type compressor
977. If a drum of dichlorodifluoromethane partially filled with liquid refrigerant is stored in a 32° room, the pressure in the drum would be:
   a) 10 lbs
   b) 124 lbs
   c) 16 lbs
   d) 30 lbs
978. A reed valve normally has:
   a) 4 springs
   b) 2 springs
   c) no springs
   d) the number of springs is determined by the valve size
979. The automatic expansion valve is controlled by:
   a) evaporator temperature
   b) evaporator coil superheat
   c) the rate of refrigerant condensation
   d) evaporator pressure
980. The thermostatic expansion valve has which three operation pressures?
   a) evaporator pressure, spring pressure and suction pressure
   b) evaporator pressure, bulb pressure and condensing pressure
   c) evaporator pressure, spring pressure and political pressure
   d) evaporator pressure, spring pressure and bulb pressure
981. The thermostatic expansion valve is controlled by:
   a) coil temperature
   b) the difference in gas temperature and the temperature corresponding to the gas pressure
   c) the difference in gas pressure and the pressure corresponding to the gas temperature
   d) coil pressure
982. When the pressure difference across a capillary tube is increased:
   a) the flow of refrigerant increases
   b) the flow of refrigerant decreases
   c) it gets noisy
   d) there is no change in flow rate
983. The touching surfaces between a thermostatic expansion valve bulb and the suction line must be:
   a) clean and bright
   b) clean and tight

c) downstream of the equalizing line
   d) above the coil
984. Flash gas is:
   a) useful in cooling the product
   b) necessary in metering devices
   c) not necessary
   d) greater in automatic devices
985. The "heat of compression" is:
   a) carried away in the evaporator
   b) sometimes wasted
   c) less on water-cooled compressors
   d) carried away in the cooling medium leaving the condenser
986. The two basic types of evaporators are:
   a) finned and plate
   b) prime surface and finned
   c) direct expansion feed and plate
   d) flooded and direct expansion feed
987. Natural draft condensers are most frequently found in:
   a) residential cooling
   b) large industrial plants
   c) household refrigerators
   d) room air conditioners
988. Liquid slugging is:
   a) the pounding of liquid refrigerant in the suction line at a point of restriction
   b) a presence of liquid in the condenser causing excessive noise
   c) liquid in compressor clearance space
   d) excessive liquid refrigerant in the receiver
989. All refrigeration compressor valves are opened and closed by:
   a) external springs
   b) inherent spring tension
   c) pressure difference
   d) a camshaft
990. The horizontal shell and tube condenser is more efficient than the vertical shell and tube condenser because:
   a) it is mounted in a horizontal position
   b) it is easier to clean
   c) the water passes through it more than once
   d) it takes advantage of more air cooling
991. The capacity of an evaporative condenser depends on the:
   a) fan horsepower
   b) amount of heat that exiting air is capable of absorbing
   c) entering air wet bulb
   d) temperature of the entering air

992. Prime surface evaporators operating below 32°F must frequently be:
    a) cleaned
    b) purged
    c) painted
    d) defrosted
993. Secondary refrigerants are usually:
    a) Prestone
    b) alcohol
    c) water or brine
    d) expensive
994. Desired fresh food storage temperature range is:
    a) 0 to 20°F
    b) 20 to 30°F
    c) 35 to 45°F
    d) 50 to 60°F
995. Frozen food storage temperature is usually:
    a) -40 to -60°F
    b) below 32°F
    c) 0 to -10°F
    d) 35 to 45°F
996. For comfort cooling, the temperature is usually maintained at what degree below the ambient temperature?
    a) 5°F
    b) 50 to 60°F
    c) 10 to 12°F
    d) 2 to 3°F
997. A screen is usually located at the:
    a) valve outlet
    b) valve seat
    c) bellows section
    d) valve inlet
998. An under-capacity valve:
    a) tends to flood the evaporator
    b) tends to starve the evaporator
    c) will work well at heavy loads
    d) none of the above
999. Valve seats are usually made of:
    a) the same materials as the needles
    b) a material that can be adjusted
    c) softer material than the needles
    d) harder material than the needles
1000. Automatic expansion valves used to replace cap tubes:
    a) are the simplest and safest way to replace a cap tube
    b) no provision for pressure difference is needed

c) none of these answers is correct
d) must have a high starting torque motor compressor

1001. A leaking needle and seat on an automatic expansion valve will:
    a) flood over evaporator if a receiver is used
    b) not possibly admit liquid to compressor
    c) not react the same as a bleeder or by-pass type
    d) have no effect on the system at all

1002. Moisture in an automatic expansion valve system:
    a) does not affect this system at all
    b) acts the same whether compressor is running or not
    c) does not cause a restriction in the system
    d) acts the same as a plugged screen

1003. To raise the temperature on an automatic expansion valve system:
    a) lessen spring tension
    b) increase spring tension
    c) increase running time on compressor
    d) does not require manual adjustment

1004. In adjusting an automatic expansion valve:
    a) less tension on spring admits more refrigerant to the evaporator
    b) less tension on spring admits less refrigerant to the evaporator
    c) temperature of evaporator will not change
    d) will cycle compressor more often

1005. A system using an AXV is cycled by the:
    a) float level
    b) low-side pressure
    c) thermostat
    d) pressure control

1006. In the space of 2 ft x 4 ft x 8 ft, there is:
    a) 14 ft^2
    b) 14 ft^3
    c) 64 ft^2
    d) 64 ft^3

1007. A 100 mesh screen has:
    a) 10 openings/in^2
    b) 1000 openings/in^2
    c) 100 openings/in^2
    d) 50 openings/in^2

1008. An automatic expansion valve is operated by:
    a) high-side pressure
    b) low-side pressure
    c) liquid live pressure
    d) atmospheric pressure

1009. The pressure of R-12 at 70°F would be approximately:
    a) 65 lbs

b) 70 lbs
c) 78 lbs

1010. What happens when a capillary tube system is overcharged?
   a) nothing
   b) it has a low head pressure
   c) it will partially defrost
   d) it will sweat and frost back

1011. Where is the most common place to install a drier in the system?
   a) in the liquid line
   b) between the compressor and the condenser
   c) between the condenser and the liquid line
   d) between the refrigerant control and the evaporator

1012. Where is the liquid receiver usually located in a TEV system?
   a) there is none
   b) in the evaporator
   c) at the outlet of the condenser
   d) in the suction line

1013. Is there any oil in the condenser?
   a) yes
   b) only if the system is water-cooled
   c) only if too much oil is in the system
   d) no

1014. How can you tell when a drier without an indicator is absorbing moisture?
   a) the drier will become cold when the unit is operating
   b) the drier will become warm when the unit is operating
   c) the head pressure will rise
   d) the head pressure will fall

1015. A strainer screen is supposed to remove:
   a) solid impurities
   b) acid
   c) oil
   d) moisture

1016. If a resistance is 10 ohms and the voltage is 230 volts, what is the current in amperes?
   a) 10
   b) 23
   c) 100
   d) 15

1017. Air-cooled condensers usually handle how much air as compared to the evaporator coil?
   a) more
   b) less
   c) the same

1018. On a belt-driven blower, if the size of the motor pulley is decreased, the speed of the blower will:
   a) increase
   b) decrease
   c) remain the same
1019. What kind of compressor motor uses one run capacitor and no relay?
   a) PSC
   b) CS
   c) CSR
1020. What kind of compressor motor uses two start capacitors and one run capacitor?
   a) PSC
   b) CS
   c) CSR
1021. If the voltage is 120 and the current is 13 amps, the resistance is how many ohms?
   a) 12.2
   b) 10.04
   c) 9.23
   d) 7.11
1022. The watt input per ton of cooling is _____ for air-cooled units than for water-cooled units.
   a) higher
   b) lower
   c) the same
1023. The cfm per ton on a packaged air conditioner is normally:
   a) 300
   b) 400
   c) 500
   d) 600
1024. The sensible heat ratio on most packaged air conditioners is approximately:
   a) 50
   b) 60
   c) 75
   d) 95
1025. Automatic expansion valve systems are usually:
   a) dry
   b) flooded
   c) very large tonnage
   d) the same as high-side floats
1026. The capacitance formula is: 2650 times the amps divided by the volts equals the mfd.
   a) true
   b) false

1027. Refrigerant 502 would be in Group II.
   a) true
   b) false
1028. Methyl formate would be in Group I.
   a) true
   b) false
1029. Refrigerant 30 would be in Group I.
   a) true
   b) false
1030. Methyl chloride would be in Group II.
   a) true
   b) false
1031. Sulphur dioxide would be in Group III.
   a) true
   b) false
1032. Refrigerant 500 would be in Group I.
   a) true
   b) false
1033. Ethyl chloride would be in Group II.
   a) true
   b) false
1034. A control circuit must be low voltage.
   a) true
   b) false
1035. The power input kWh and horsepower per ton rating for an air-cooled condensing unit is higher than for a water-cooled condensing unit.
   a) true
   b) false
1036. The operating cost of an air-cooled condensing unit is higher than the operating cost of a water-cooled condensing unit using city water under comparable conditions.
   a) true
   b) false
1037. The reason for using air-cooled condensing units in preference to water-cooled condensing units is their more attractive appearance.
   a) true
   b) false
1038. In figuring operating costs, the running time of the unit is used in the calculations, not the full load hours.
   a) true
   b) false
1039. The water cost of a cooling tower is due to "bleed off" and the evaporation of water.

a) true
b) false

1040. Air-cooled condensing units located outside the building usually use a propeller type fan.
a) true
b) false

1041. The components of an air-cooled condensing unit include the thermal expansion valves or metering device and the evaporator coil.
a) true
b) false

1042. Refrigerant 13 would be in Group I.
a) true
b) false

1043. Refrigerant 11 would be in Group I.
a) true
b) false

1044. The maximum permissible quantities of Group II refrigerants for indirect systems are: commercial, 600 lbs; residential, 300 lbs.
a) true
b) false

1045. Sweat joints on copper tubing used in refrigerating systems containing Group II or Group III shall be soldered joints.
a) true
b) false

1046. Refrigerant pipe joints erected on the premises shall be exposed for visual inspection prior to being covered or enclosed.
a) true
b) false

1047. Refrigerant piping crossing an open space which affords passageway in any building shall be not less than 7½ feet above the floor unless against the ceiling of such space.
a) true
b) false

1048. Aluminum, zinc or magnesium shall not be used in contact with methyl chloride in a refrigerating system.
a) true
b) false

1049. Magnesium alloys shall not be used in contact with any halogenated refrigerant.
a) true
b) false

1050. Discharge of pressure-relief devices and fusible plugs on all systems containing more than 6 lbs of Group II or Group III refrigerants shall be to

the outside of the unit.
a) true
b) false

1051. A brazed joint is a gas tight joint obtained by the joining of metal parts at temperatures higher than 1000°F but less than the melting temperatures of the joined parts.
a) true
b) false

1052. Brine is any liquid used for the exchange of heat without a change in its state.
a) true
b) false

1053. Companion (or block) valves are pairs of mating stop valves.
a) true
b) false

1054. A compressor is a specific machine with or without accessories for a given refrigerant vapor.
a) true
b) false

1055. A compressor unit is a condensing unit less the condenser and liquid receiver.
a) true
b) false

1056. A condenser is a vessel or arrangement of pipe or tubing which is used for the holding of gas.
a) true
b) false

1057. A compressor cannot be "unloaded" by throttling the flow of suction gas.
a) true
b) false

1058. A broken discharge reed can be caused by high head pressure.
a) true
b) false

1059. A loose rod bearing will always result in a compressor motor burnout.
a) true
b) false

1060. The length of the suction line can be the cause of a compressor motor burnout.
a) true
b) false

1061. A worn wrist pin causes the suction pressure to run higher than it normally would.
a) true
b) false

1062. A leaking discharge reed will prevent a compressor from pulling down to as low a vacuum as it normally would.
   a) true
   b) false
1063. A leaking suction reed will prevent a compressor from pulling as deep a vacuum as it normally would.
   a) true
   b) false
1064. The higher the suction pressure, the higher the compressor capacity.
   a) true
   b) false
1065. Copper plating is much more likely to occur in air conditioning compressors than in low temperature compressors.
   a) true
   b) false
1066. A suction line filter-drier is less likely to cause trouble in a low temperature compressor than in an air conditioning compressor.
   a) true
   b) false
1067. Changing an R-12 sealed unit compressor to an R-22 would increase the current draw by approximately 40%.
   a) true
   b) false
1068. A centrifugal compressor equalizes as soon as it stops turning.
   a) true
   b) false
1069. A ruptured disc which will blow with high head pressure to by-pass from the high-side back into the low-side is an excellent protective device to guard the compressor against abnormal head pressure.
   a) true
   b) false
1070. A direct drive compressor must have not more than .010" misalignment between compressor shaft and motor shaft to prevent damage to the main bearings, seal and couplings.
   a) true
   b) false
1071. The lower the superheat of the suction gas, the less efficient the compressor.
   a) true
   b) false
1072. The dome of a Scotch yoke compressor is on the high pressure side of the system.
   a) true
   b) false

1073. Acid or moisture can be detected in a sealed unit compressor by using a megohmeter.
   a) true
   b) false
1074. Ammonia compressors cannot be used with the other refrigerants.
   a) true
   b) false
1075. If a compressor uses individual cylinder unloading, the unloading is always accomplished by holding the suction valve open by the force of a spring.
   a) true
   b) false
1076. Moisture indicators are to check the amount of frost.
   a) true
   b) false
1077. The pour point of oil is important.
   a) true
   b) false
1078. Crankcase heaters are for heating oil.
   a) true
   b) false
1079. A sight glass shows flow of refrigerant.
   a) true
   b) false
1080. A system would be safe if it were evacuated to a temperature a little lower than evaporator temperature.
   a) true
   b) false
1081. Carbon Dioxide would be in Group III.
   a) true
   b) false
1082. Butane would be in Group II.
   a) true
   b) false
1083. Ethane would be in Group II.
   a) true
   b) false
1084. Ethylene would be in Group I.
   a) true
   b) false
1085. Isobutane would be in Group III.
   a) true
   b) false

1086. Propane would be in Group II.
   a) true
   b) false
1087. Dichloroethylene would be in Group I.
   a) true
   b) false
1088. Ammonia discharge (where ammonia is used) may be into a tank of water which shall be used for no purpose except ammonia absorption.
   a) true
   b) false
1089. Sulphur dioxide discharge (where sulphur dioxide is used) may be into a tank of absorptive brine.
   a) true
   b) false
1090. Pressure vessels over 3 ft^3 but less than 10 ft^3 have no relief device.
   a) true
   b) false
1091. A shaft seal is necessary.
   a) true
   b) false
1092. The temperature of a gas vapor is raised by applying pressure.
   a) true
   b) false
1093. Solvent is generally used to clean up a system.
   a) true
   b) false
1094. R-12 refrigerant comes in a white color-coded cylinder.
   a) true
   b) false
1095. A reciprocating compressor can be driven by direct drive.
   a) true
   b) false
1096. The two basic types of compressors are open and closed.
   a) true
   b) false
1097. A system may be no more than 8.6% wet.
   a) true
   b) false
1098. Moisture cannot affect the operation of a metering device as long as there is no more than 8.6% water in the system.
   a) true
   b) false

1099. A hot gas line is for holding the compressor's refrigeration.
   a) true
   b) false
1100. Copper tubing is connected to a refrigeration system by using 95/5 solder.
   a) true
   b) false
1101. Hot gas from a condensing unit can be used to defrost an evaporator.
   a) true
   b) false
1102. Moisture would have no effect on a system operating above freezing (such as an air conditioner with a 45° evaporator).
   a) true
   b) false
1103. Ordinary copper tube differs from refrigeration grade tube.
   a) true
   b) false
1104. The position of the muffler is important.
   a) true
   b) false
1105. Design working pressure is the maximum allowable working pressure for which a specific part of a system is designed.
   a) true
   b) false
1106. An evaporator is that part of the system in which liquid refrigerant is vaporized to produce refrigeration.
   a) true
   b) false
1107. A fusible plug is a device for units over 7½ tons.
   a) true
   b) false
1108. An expansion coil is an evaporator constructed of pipe or tubing.
   a) true
   b) false
1109. "High-side" means the very top of the compressor.
   a) true
   b) false
1110. A liquid receiver is for the storage of compressor oil.
   a) true
   b) false
1111. "Low-side" means the parts on the bottom of the compressor.
   a) true
   b) false

1112. A non-positive displacement compressor is a compressor in which increase in vapor pressure is attained without changing the internal volume of the compression chamber.
   a) true
   b) false
1113. A positive displacement compressor is a compressor in which increase in vapor pressure is attained by changing the internal volume of compression.
   a) true
   b) false
1114. A pressure-imposing element is any device or portion of the equipment used for the purpose of increasing the oil pressure.
   a) true
   b) false
1115. A pressure limiting device is a pressure-responsive mechanism designed to automatically increase the refrigerant vapor pressure.
   a) true
   b) false
1116. A pressure relief device is a pressure-actuated safety valve.
   a) true
   b) false
1117. Monochlorodifluoromethane is the word for F-12.
   a) true
   b) false
1118. Dichlorodifluoromethane is the word for F-22.
   a) true
   b) false
1119. Ammonia would be in Group III.
   a) true
   b) false
1120. Methylene chloride would be in Group I.
   a) true
   b) false
1121. Refrigerant 22 would be in Group II.
   a) true
   b) false
1122. Refrigerant 12 would be in Group I.
   a) true
   b) false
1123. Unloading a cylinder in a compressor increases the head pressure.
   a) true
   b) false
1124. By-passing some of the discharge gas back into the suction line decreases the discharge temperature.

a) true
b) false

1125. Liquid refrigerant returning with the suction gas prevents proper lubrication.
a) true
b) false

1126. Low suction pressure can cause a sealed unit motor to burn out.
a) true
b) false

1127. Liquid refrigerant in the crankcase will not be harmful in an enclosed crankcase compressor.
a) true
b) false

1128. The greatest problem in two-stage systems using two compressors with R-12 is overheating.
a) true
b) false

1129. A splash-lubricated compressor can turn in only one direction.
a) true
b) false

1130. Dropping the suction pressure on a straight air-cooled system, a standard system with a cooling tower, or on an evaporative condenser system will drop the head pressure.
a) true
b) false

1131. Dropping the suction pressure lowers the current draw on any refrigeration system.
a) true
b) false

1132. Dropping the suction pressure lowers the compressor discharge temperature on any type of system.
a) true
b) false

1133. A muffler will often be more effective in preventing vibration of the refrigerant lines than will a bellows type vibration eliminator.
a) true
b) false

1134. Vibration eliminator fittings should always be installed parallel to the compressor crankshaft.
a) true
b) false

1135. A compressor with a leaky discharge valve may "backlash" when it stops.
a) true
b) false

1136. A compressor which must have the belts pulled up so tightly to prevent slippage so that no "bow" is seen in the off side of the belts when running does not have enough grooves in the pulleys.
   a) true
   b) false
1137. A system does not need to follow normal rules of piping for proper oil return if a high-side oil separator is installed.
   a) true
   b) false
1138. Pulling head bolts down tighter on one head of a V-type compressor than on the other head can damage the compressor.
   a) true
   b) false
1139. Bolting a compressor firmly to a heavy concrete block will prevent vibration and will result in longer compressor life.
   a) true
   b) false
1140. The advantages of using a centrifugal fan on an air-cooled condensing unit are quieter operation and allowing the installation of air ducts.
   a) true
   b) false
1141. The preferable location of the receiver on an air-cooled condensing unit is below the level of the condenser-coil outlet.
   a) true
   b) false
1142. On most remote air-cooled condensing units, the condenser blower runs all the time and the compressor cycles on head pressure.
   a) true
   b) false
1143. A good pressure setting for the relief valve on an air-cooled condensing unit using R-22 is 300 psi where the normal operating condensing temperature does not exceed 125°F.
   a) true
   b) false
1144. It is reasonable to expect that the running power input in kilowatts for an air-cooled condensing unit used for air conditioning will not exceed ½ kW per ton.
   a) true
   b) false
1145. On an air-cooled, self-contained air conditioning unit, the cfm handled by the condenser fan is often double or higher than the cfm handled by the evaporator.
   a) true
   b) false

1146. According to ASHRAE, the pressure drop for the refrigerant in suction lines should not exceed 2 psi.
   a) true
   b) false
1147. According to ASHRAE, the pressure drop in refrigerant liquid lines should not exceed 4 psi.
   a) true
   b) false
1148. The voltage rating of a capacitor indicates the minimum rating at which it should be used.
   a) true
   b) false
1149. An evaporative condenser can handle the gas discharge from:
   a) one compressor only
   b) centrifugal compressors only
   c) several compressors, providing it has the capacity
1150. When "hot gas" enters the evaporator, it:
   a) becomes warmer
   b) condenses
   c) cools
   d) evaporates
1151. In a low-side float valve, the float ball and mechanism are:
   a) in the high-pressure side of the system
   b) in the low-pressure side of the system
   c) completely outside the system
1152. The function of an oil regenerator is to:
   a) return oil carry-over back to the regenerator
   b) purify the oil
   c) separate the refrigerant from the oil
1153. For what application is a snap-action two-temperature valve usually used?
   a) where a small temperature difference is desired
   b) where a large temperature difference is desired
   c) in small and large cabinet combination installations
   d) anywhere in a multiple installation
1154. Where a cooling pond is used, the water is cooled primarily by:
   a) evaporation
   b) convection
   c) radiation
1155. The expansion valve most adaptable for varying load conditions is the:
   a) thermostatic valve
   b) automatic expansion valve
   c) high-side float

1156. In all condenser operations, the heat gained by the condensing medium must equal:
   a) the amount of refrigerant condensed
   b) the heat given up by the refrigerant
   c) the amount of water used
1157. Superheat vapor is:
   a) wet vapor at its condensing temperature
   b) dry vapor at its boiling point
   c) vapor which has been heated above its boiling point
1158. The actual amount of heat removed from the heat load by a refrigerating system is known as the:
   a) net refrigerating effect
   b) volumetric efficiency
   c) compression ratio
1159. Too much refrigerant will cause the most trouble in a:
   a) low-side float system
   b) thermostatic expansion valve system
   c) high-side float system
1160. An increase in the heat load will cause the thermostatic expansion valve to:
   a) increase the flow of refrigerant
   b) decrease the flow of refrigerant
   c) balance
1161. The compressor compresses the vapor to:
   a) raise its superheat
   b) raise its pressure and temperature
   c) add the heat of compression
1162. An evaporative condenser:
   a) is always used where there is a water shortage
   b) must always be located outdoors
   c) is a combination of an air-cooled condenser, a water-cooled condenser and a cooling tower
1163. Pressure is defined as:
   a) 14.7 psi
   b) weight of the air above us
   c) force per unit area
1164. A low-side float is used on systems having:
   a) a critical charge
   b) one evaporator
   c) one or more evaporators
1165. The purpose of the heat exchanger is to:
   a) prevent sweating of the suction line
   b) subcool the liquid
   c) decrease the amount of superheat

1166. The liquid refrigerant evaporated in cooling the rest of the liquid as it goes through the metering device is known as:
   a) superheat
   b) subcooling
   c) flash gas
1167. When installing a capillary tube, the only adjustment is its:
   a) superheat
   b) diameter
   c) length
1168. F-12 vapor enters the condenser at 117.4 psig and 100°F. If it leaves the condenser at 85°F, its pressure is:
   a) 91.9 psig
   b) 117.4 psig
   c) 154 psig
1169. Unlike the vertical shell and tube condenser, the water in the horizontal shell and tube condenser:
   a) makes more than one pass through the condenser
   b) is recirculated
   c) is treated
1170. What size motor-compressor is used as a replacement?
   a) one size larger
   b) one size smaller
   c) 1/20 hp more
   d) the same size
1171. An excessive pressure drop in an evaporator will cause a thermostatic expansion valve to:
   a) operate at a higher temperature
   b) operate at a lower temperature
   c) equalize
1172. Temperature is defined as the:
   a) measurement of the amount of heat
   b) amount of heat required to cause a change of state
   c) measurement of the intensity of heat
1173. In condenser operations, the lower the condensing temperature, the:
   a) less water will be used
   b) less efficient the compressor becomes
   c) lower the condensing pressure
1174. Vapor which is in contact with the liquid from which it is evaporated, or is at its evaporating temperature, is:
   a) superheated vapor
   b) saturated vapor
   c) dry vapor
1175. The temperature of the medium being cooled must be:
   a) below evaporator temperature

b) equal to evaporator temperature
c) above evaporator temperature

1176. An under-charged high-side float system will:
    a) operate with a starved evaporator
    b) pass vapor
    c) remain closed

1177. The low-side float closes with a:
    a) rise in the liquid level
    b) drop in the liquid level
    c) rise in the evaporator temperature

1178. The capacity of an evaporative condenser depends on the:
    a) temperature of the entering air
    b) wet bulb temperature of entering air
    c) amount of water circulated

1179. If 970 Btu were added to a pound of water at a temperature of 212°F at atmospheric pressure, the temperature of the steam would be:
    a) 1180°F
    b) 212°F
    c) 758°F

1180. The two basic types of evaporators are:
    a) bare pipe and finned
    b) flooded and dry
    c) blower coil and gravity

1181. A capillary tube system has:
    a) a condenser receiver
    b) no receiver
    c) a receiver

1182. The automatic expansion valve tries to maintain:
    a) the evaporator full of liquid
    b) a constant pressure
    c) the flow of refrigerant according to the load

1183. $H=WS(t_2-t_1)$ If the specific heat of beef is 0.77, the amount of heat to be removed from 100 lbs of beef in cooling it down from 75° to 35° is:
    a) 7,700 Btu
    b) 400 Btu
    c) 3,080 Btu

1184. The purpose of the equalizing line between the condenser and the receiver is to:
    a) compensate for pressure drop
    b) prevent flash gas
    c) balance pressure

1185. When the pressure on a liquid is increased, its boiling point:
    a) is lowered

b) remains the same
c) is raised

1186. When the compression ratio of the compressor increases, the capacity of the compressor:
   a) increases
   b) decreases
   c) remains the same

1187. A shortage of refrigerant will cause a low-side float to:
   a) remain closed
   b) operate at a lower suction pressure
   c) remain open

1188. A slightly over-charged low-side float system will:
   a) cause flooding
   b) cause the float to stay open
   c) have no effect on the evaporator

1189. The amount of heat required to change a liquid to a vapor with no change in temperature is called latent heat of:
   a) fusion
   b) vaporization
   c) sublimation

1190. In a counterflow condenser, the coldest water:
   a) meets the coldest refrigerant
   b) meets the warmest refrigerant
   c) enters at the top

1191. Latent heat of fusion is the amount of heat required to:
   a) change a liquid to a solid with no change in temperature
   b) change a solid to a vapor
   c) raise the temperature of a substance above its boiling point

1192. A higher heat transfer per ft^2 of evaporator is obtained with a:
   a) dry expansion system
   b) flooded evaporator
   c) finned evaporator

1193. A high-side float is used on a system having:
   a) any number of evaporators
   b) one evaporator
   c) two evaporators of different temperatures

1194. The thermostatic expansion valve controls the:
   a) temperature of the evaporator
   b) pressure in the evaporator
   c) flow of refrigerant to the evaporator

1195. The Btu required to raise the temperature of 10 lbs of water 300°F is:
   a) 30 Btu
   b) 40 Btu
   c) 3000 Btu

1196. Sub-cooling of the liquid refrigerant is desirable because:
   a) it prevents liquid from returning to the compressor
   b) it reduces the amount of flash gas
   c) the compressor operates at a lower head pressure
1197. 746 watts of electrical energy or 2,545 Btu/hour of heat energy is equal to:
   a) 1 ton of refrigeration
   b) 1 horsepower
   c) 1 kilowatt
1198. With a high suction pressure, the capacity of the compressor is:
   a) not changed
   b) decreased
   c) increased
1199. An externally-equalized thermostatic expansion valve:
   a) maintains constant evaporator pressure
   b) keeps the evaporator filled with liquid
   c) overcomes pressure drop in an evaporator
1200. A rise in the liquid level of a high-side float will cause:
   a) the float to open
   b) the float to close
   c) floodback
1201. One must be careful when handling the old compressor oil:
   a) to keep it clean
   b) because it may be hot
   c) because it may be acidic
   d) because it may spill
1202. How many kinds of capacitors are there?
   a) one
   b) two
   c) three
   d) four
1203. When installing a water regulating valve on a refrigeration system, it should be located:
   a) on the water inlet side
   b) on the water outlet side
   c) either at inlet or outlet side
1204. What type of capacitor is usually used for continuous operation?
   a) the run capacitor
   b) any type
   c) copper graphite
   d) an oversize capacitor
1205. The normal cfm/ton of cooling for residential units is:
   a) 200 cfm/ton
   b) 400 cfm/ton
   c) 650 cfm/ton

1206. In case of condensing water failure, which of the following safety devices should operate first?
   a) high-pressure cut-out switch
   b) low water cut-out switch
   c) pressure relief valve
1207. How is the heating element in a butter conditioner connected into the system?
   a) parallel with the other accessories
   b) in series with the other accessories
   c) independent of the motor
   d) none of the above
1208. How is the starting capacitor connected to the start winding electrically?
   a) in series
   b) in parallel
   c) either series or parallel
   d) between the start and run winding
1209. A saturated vapor is a vapor:
   a) saturated with liquid droplets
   b) at a temperature corresponding to the pressure
   c) at a temperature higher than 5°
1210. What is another name for a capacitor?
   a) meter
   b) stator
   c) condenser
   d) rotor
1211. How are the light switch and light wired?
   a) in series with the motor
   b) in parallel with the motor
   c) independent of the motor
   d) in series with the thermostat
1212. What is the purpose of the freezer door heater?
   a) to prevent condensation from forming and freezing
   b) to provide enough heat to give an adequate seal to the door
   c) to keep the gasket pliable
   d) to prevent moisture from dripping on the floor
1213. How many circuits are possible if two wires are fastened to one terminal?
   a) one
   b) two
   c) three
   d) four
1214. What oil should be used on bronze motor bearings?
   a) SAE 0 to 10
   b) SAE 10 to 30
   c) SAE 30 to 50

d) SAE 50 to 60
1215. If the condensing water flow stops, which of the following safety devices should be the second to go into operation in event of failure of the first?
a) high-pressure cut-out switch
b) low water cut-out switch
c) pressure relief valve
1216. While changing to ice, water:
a) does not change in volume
b) contracts
c) expands
1217. What should be done if a fan blade is bent out of line?
a) straighten it
b) replace it
c) operate it "as is"
d) if possible, bend the other blades to match
1218. The water in an evaporative condenser is cooled chiefly by:
a) conduction
b) evaporation
c) convection
1219. If a gas mask is discovered with a broken seal on the canister:
a) the seal should be renewed
b) the canister should be renewed after a two-year period
c) the canister should be renewed at once
1220. At what part of the refrigerating cycle is the motor current the highest?
a) the instant of starting
b) the time that the condenser pressure is the highest
c) the time that the low-side pressure is the highest
d) the time that the low-side pressure is the lowest
1221. What is one advantage of an oil separator?
a) keeps the oil in the compressor
b) low cost
c) easy to service
d) traps the moisture
1222. If the oil separator float collapses:
a) nothing will happen
b) the valve will stay open
c) the valve will stay closed
d) the liquid refrigerant will short circuit into the crankcase
1223. An oil separator must be mounted:
a) in any position
b) level
c) suspended from the condenser line
d) below the compressor

1224. What could be wrong if a voltage exists between the motor-compressor housing and a water pipe when the power is turned on?
   a) an open circuit
   b) a closed circuit
   c) a short
   d) a ground
1225. If the substance changes from a solid directly into a gas without liquefying, this process is referred to as:
   a) solidification
   b) sublimation
   c) condensation
1226. In a large ice-making plant, the effect of any re-adjustment of the expansion valve would:
   a) not be noticeable
   b) not be immediately noticed
   c) be immediately noticed
1227. Where should the check valve be located when used in two-temperature installations?
   a) in the warmest coil suction line
   b) in the coldest coil suction line
   c) in the liquid line
   d) at the compressor
1228. The canisters (or cartridges) for gas masks, if not used, should be renewed every:
   a) two years
   b) three years
   c) five years
1229. The two thermometers of a sling psychrometer read alike when the relative humidity is:
   a) 0 %
   b) 50 %
   c) 100 %
1230. A compound gauge, if provided for a positive displacement compression system, should be located on the:
   a) high-side of the system
   b) low-side of the system
   c) receiver
1231. A lower condenser pressure will require:
   a) greater work for the compressor
   b) less work for the compressor
   c) greater power to drive the compressor
1232. At what spot would moisture freeze in a window unit using a capillary tube?
   a) on the low-side
   b) on the screen

c) no place
d) in the evaporator
1233. Lower head pressures and temperatures can be obtained by use of:
   a) compound compression
   b) single-stage compression
   c) evaporative condensers
1234. Leakage at the shaft seal of an R-11 centrifugal compressor is usually:
   a) of little importance to efficient operation
   b) inward to compressor
   c) outward to the compressor
1235. Where does the air go after it cools the condenser?
   a) indoors
   b) outdoors
   c) into the evaporator
   d) into the fresh air intake
1236. Why must window air conditioner units be leveled?
   a) to provide proper air circulation
   b) to provide for proper flow of the condensate
   c) to prevent freeze-up of the evaporator
   d) to compensate for wind direction
1237. Why is it important to mount the unit in a level position?
   a) for lubrication
   b) to allow good refrigerant flow
   c) to minimize noise
   d) to insure proper control of condensate
1238. In laying-up a plant for winter months, it is considered good practice to store the refrigerant in the:
   a) accumulator
   b) evaporator
   c) condenser and/or the receiver
1239. Where does the condenser cooling air come from?
   a) indoors
   b) outdoors
   c) from the cooling unit
   d) from outdoors only in cool weather
1240. Adding more calcium chloride to the brine will:
   a) lower its freezing point under certain conditions
   b) always lower its freezing point
   c) have no effect on its freezing point
1241. If the temperature of a cooling coil is below the dew point of the air passing through it, the:
   a) air will be free of any moisture
   b) air will evaporate any moisture on the cooling coil
   c) moisture will condense out of the air

1242. If the unit rumbles as it stops, what may be wrong?
   a) nothing
   b) the wall is not strong enough
   c) the filters are clogged
   d) the unit may not be mounted level
1243. Of the conditions listed below, the one having the most adverse effect on the proper operation of a low-side float would be:
   a) too much refrigerant in the system
   b) the correct amount of refrigerant in the system
   c) not enough refrigerant in the system
1244. A discharge line oil trap or separator is utilized in a refrigeration system having a:
   a) high volumetric efficiency
   b) centrifugal compressor
   c) reciprocating compressor
1245. Which type of evaporating coil is most used on chest type freezers?
   a) shell liner
   b) plate on shelf
   c) open coils
   d) flooded
1246. What is the recommended temperature range for a chest type food freezer?
   a) 0°F to 10°F
   b) -20°F to -10°F
   c) -30°F to -25°F
   d) -40°F to -30°F
1247. Dirty filters in an air conditioning system cause:
   a) high suction pressure
   b) high discharge pressure
   c) low suction pressure
   d) none of the above
1248. How is the defrost water kept from freezing in the drain pan and drain tubes?
   a) there is no danger of freezing
   b) it drains away too fast to freeze
   c) by the heat in the liquid line
   d) by electrical resistance heaters
1249. In the evaporative condenser, the tubes are usually:
   a) finned
   b) bare
   c) studded
1250. What is the purpose of the relay?
   a) to protect against an overload
   b) to keep the motor from running too fast
   c) to disconnect the starting winding at the proper moment
   d) to take the place of the thermostat

1251. What instrument should be used to check the receptable outlet?
    a) ohmmeter
    b) megohmmeter
    c) voltmeter
    d) ammeter
1252. How many terminals are used on most hermetic units which operate at 3,400 rpm?
    a) three
    b) four
    c) two
    d) five
1253. What may be housed in the junction box on some units?
    a) relay
    b) thermostat
    c) mullion heater
    d) defrost control
1254. What instrument should be used to check for a short circuit?
    a) a voltmeter
    b) an ohmmeter
    c) an ammeter
    d) a wattmeter
1255. How many sets of points does a hot wire relay have?
    a) one
    b) two
    c) three
    d) four
1256. A relay is usually located:
    a) in the thermostat
    b) in the motor
    c) on the condensing unit
    d) in the capacitor box
1257. What will most thermal relays do when the unit is using too much current?
    a) nothing
    b) burn out
    c) open the circuit
    d) close the starting circuit
1258. A refrigerant that is flammable is:
    a) methyl chloride
    b) sulphur dioxide
    c) carbon dioxide
1259. An accepted method of preventing frost from forming on blast freezers is to use:
    a) hot gas
    b) electric heaters

c) brine spray
1260. A Freon 11 system generally utilizes a:
   a) reciprocating compressor
   b) rotary compressor
   c) centrifugal compressor
1261. How accurately must a dehumidifier with a capillary tube be charged?
   a) plus or minus 1 lb
   b) with an extra ½ lb
   c) very accurately
   d) until it frosts back, and then purge out a pound
1262. The pH of brine may be increased by adding:
   a) caustic soda
   b) carbon dioxide gas
   c) sodium chloride
1263. What type of cycling device is often used on a dehumidifier?
   a) pressure motor control
   b) humidistat
   c) rheostat
   d) high-side cut-off
1264. What happens to the pressures in a capillary tube system during the off cycle?
   a) they both decrease a little
   b) the both increase a little
   c) the low-side pressure increases and the head pressure stays constant
   d) the pressures tend to equalize
1265. Of the substances listed below, which is a chlorinated hydrocarbon?
   a) carbon dioxide
   b) ammonia
   c) water
   d) freon
1266. Superheated Freon 12:
   a) must contain liquid globules
   b) may or may not contain liquid globules
   c) must be completely free from liquid globules
1267. A low-side float valve will maintain a:
   a) liquid level in the evaporator which will correspond to load and pressure
   b) constant liquid level in evaporator regardless of load and pressure
   c) constant pressure in the evaporator
1268. What may be wrong if the system frosts back?
   a) undercharged
   b) overcharged
   c) inefficient compressor
   d) moisture in the system
1269. How is the defrost disposed of in an upright, frost-free freezer?
   a) out a normal drain

b) evaporated in the motor compressor condenser compartment
c) evaporated by a "hot gas" defrost system
d) no frost accumulates

1270. A liquid whose boiling point is above 32°F could:
    a) not be used as a refrigerant
    b) be used as a refrigerant
    c) be used only with Freon

1271. The addition of oil to refrigerant R-12:
    a) raises its boiling point
    b) lowers its boiling point
    c) has no effect on its boiling point

1272. Why must relay covers be kept as tight as possible?
    a) it is not necessary
    b) dust will cause the points to deteriorate
    c) to keep the unit cool
    d) to keep the unit warm

1273. Chlorine is used in water treatment to:
    a) kill micro-organisms
    b) reduce inorganic matter
    c) increase alkalinity

1274. Shell and tube type condensers are set:
    a) vertically
    b) horizontally
    c) either vertically or horizontally

1275. If an exact potential relay replacement is not available, what may be substituted for it?
    a) replace with a hot wire relay
    b) use one with a lower voltage rating
    c) use one with a higher voltage rating
    d) replace with a current relay

1276. How may an electrically grounded winding be detected?
    a) one will get a shock
    b) the unit will not run
    c) a current will flow from one terminal to another
    d) a current will flow from a terminal to the housing

1277. In condenser service, the general practice is to have the water and refrigerant circulation using the:
    a) up-flow principle
    b) parallel flow principle
    c) counter flow principle

1278. The best remedy for a squeaky V belt (rubber composition) is to:
    a) check the belt alignment
    b) lubricate the belt with oil
    c) give the belt plenty of slack

1279. Under most conditions, the best storage condition is that in which the relative humidity of the air:
   a) is well above the moisture content of the stored product
   b) is well below the moisture content of the stored product
   c) equals the moisture content of the stored product
1280. Scale formation on heat transfer surfaces (water-side of condenser tubes):
   a) may seriously hamper efficient operation
   b) should not be removed, as it retards corrosion
   c) allows an operator to operate his plant with less refrigerant
1281. Calcium chloride and lithium bromide are classed as:
   a) desaturators
   b) absorbents
   c) adsorbents
1282. To cool one gallon of milk from 75° to 45° (sp heat of milk 0.92, 1 gal. of milk weighs 8.6 lb) requires approximately:
   a) 130 Btu
   b) 240 Btu
   c) 360 Btu
1283. What circuit is energized after the motor reaches operating speed?
   a) both starting and running windings
   b) only the running winding
   c) only the starting winding
   d) none of them
1284. What happens to the low-side pressure when the hot gas by-pass valve opens?
   a) it stays the same
   b) it rises
   c) it lowers
   d) first it lowers, then it rises
1285. When adding a central cooling system to an existing furnace, you may have to add what controls to correlate heating and cooling functions?
   a) heat-cool thermostat and transformer
   b) heat-cool thermostat and impedance relay
   c) fan relay and pressure switch
   d) fan center and heat-cool thermostat
1286. In what position are the points in a potential relay during the off cycle?
   a) closed
   b) midway
   c) open
   d) it dependends upon the unit
1287. What is the basis of operation of the potential relay?
   a) an increase in voltage as the unit approaches and reaches its rated speed
   b) a decrease in amperage as the unit approaches and reaches its rated speed
   c) the difference in magnetic impulses

d) the potential difference between the two magnetic forces
1288. How many wires can be put under a screw terminal?
   a) 1
   b) 2
   c) 3
   d) 4
1289. How should stranded wire be connected to a screw terminal?
   a) wrap wire clockwise around screw
   b) wrap wire counter-clockwise around screw
   c) solder the wire, then wrap around the screw
   d) fasten a terminal to wire end
1290. What type of refrigerant control is in common use on small domestic freezers?
   a) capillary tube
   b) automatic expansion valve
   c) thermostatic expansion valve
   d) low-side float
1291. What is continuity?
   a) a grounded wire
   b) a broken wire
   c) a complete electrical circuit
   d) an open circuit
1292. The resistance across a clean tight terminal should be:
   a)   0 ohms
   b)   50 ohms
   c) 5,000 ohms
   d) 50,000 ohms
1293. The common lead is the:
   a) left side lead
   b) right side lead
   c) middle terminal
   d) terminal with both a running and starting winding connection
1294. How many terminals do most hermetic motors have?
   a) one
   b) two
   c) three
   d) four
1295. A two stage or compound compressor has:
   a) cylinders of equal size
   b) a larger low pressure cylinder
   c) a larger high pressure cylinder
1296. Which of the following is the largest conductor?
   a) No. 8
   b) No. 10

c) No. 12
d) No. 14
1297. The primary function of a booster compressor is to:
   a) lower the discharge pressure
   b) lower the suction pressure
   c) increase the discharge pressure
1298. What windings do most hermetic motors have?
   a) a stator and rotor winding
   b) a stator winding only
   c) two stator windings
   d) two rotor windings
1299. Compressors equipped with valve unloaders are of the:
   a) positive displacement (piston) type
   b) centrifugal type
   c) hermetic type
1300. The main function of the oil lantern in the stuffing box of compressor is to:
   a) lubricate the rod
   b) prevent gas leakage
   c) separate impurities from the oil
1301. How should a solid wire be wrapped around a terminal screw?
   a) opposite the direction the screw turns as it is tightened
   b) the same direction the screw is turned as it is tightened
   c) it should be wrapped twice around the screw
   d) it should be kept straight
1302. A steam turbine is sometimes used to drive (prime mover) a:
   a) centrifugal compressor
   b) rotary compressor
   c) reciprocating compressor
1303. Temperature expressed in degrees above absolute zero is:
   a) absolute temperature
   b) saturation temperature
   c) critical temperature
1304. Containers used for refrigerants (when charging) must be approved by:
   a) API-ASME
   b) ASRE
   c) ICC
1305. Condenser cooling water having a pH of 10 is considered:
   a) alkaline
   b) neutral
   c) acid
1306. If a refrigerator does not operate (run), what is the first circuit which should be checked?
   a) the wall outlet
   b) the thermostat motor control

c) the cabinet light circuit
d) the motor relay

1307. Tensile strength is highest in:
a) mild steel
b) aluminum
c) copper

1308. The total heat removed from the refrigerant in the condenser is:
a) normally more than that absorbed in the evaporator
b) normally less than that absorbed in the evaporator
c) the latent heat of evaporation

1309. The mechanical refrigeration system uses a _____ to circulate the refrigerant.
a) condenser
b) compressor
c) refrigerant
d) suction line

1310. The part that picks up the heat is called the:
a) unloader
b) crosshead
c) condenser
d) evaporator

1311. The part that carries the heat is called the:
a) refrigerant
b) by-pass line
c) metering device
d) motor drive device

1312. The refrigerant absorbs heat in a mechanical system when:
a) the vapor changes to a gas
b) the vapor changes to a liquid
c) it is at its critical points (temperature and pressure)
d) the liquid changes to vapor

1313. The temperature at which a change of state occurs is:
a) constant
b) the critical temperature
c) dependent on the critical pressure
d) none of the above

1314. The two most accurate methods to charge an air conditioning system with the proper charge of refrigerant are:
a) weight and temperature-pressure
b) weight and sight glass
c) frost line and temperature-pressure
d) sight glass and frost line

1315. Heat flow is from:
a) a colder body to a warmer one

b) a warmer body to a colder one
c) horizontal to vertical
d) a warm body to a hotter one

1316. In a change of state situation (vapor to liquid), if the pressure is changed:
    a) it will be the same as atmospheric pressure
    b) the vapor will not vaporize
    c) the refrigerant is useless
    d) the temperature is changed

1317. A refrigeration system must be:
    a) finger tight
    b) air wrench tight
    c) gas tight
    d) water tight

1318. The two main pressures operating in a system can be termed as:
    a) outside and inside pressure
    b) high-side and low-side pressure
    c) inside pressure and ambient pressure
    d) crankcase pressure and oil pressure

1319. To have heat transfer there must be:
    a) a temperature difference between the materials
    b) a similarity in materials
    c) no temperature difference
    d) a difference in atmospheric pressures

1320. To obtain automatic control of a refrigerator, we use a:
    a) dualstat
    b) thermostat
    c) manual switch
    d) high-pressure cut-out

1321. Reciprocating compressors consist of mainly:
    a) closely meshed gears
    b) a closely fitted piston inside of a cylinder
    c) rotating vanes moving back and forth
    d) anything that turns

1322. Most compressors are built out of:
    a) copper
    b) brass
    c) cast iron
    d) fiber glass

1323. The most common crankshaft used in compressors is a:
    a) completely straight one
    b) swash plate
    c) Scotch yoke
    d) crank throw or automotive

1324. Eccentric compressors have:
   a) extra long pistons
   b) large bearing surfaces on journals
   c) short, fat pistons
   d) small bearing surfaces on journals
1325. Small domestic compressors are usually:
   a) splash lubricated
   b) pump lubricated
   c) in sizes between 10 and 50 horsepower
   d) supplied with service valves
1326. Stationary bellows seals are used on:
   a) motors
   b) hermetic compressors
   c) open type compressors
   d) serviceable hermetic compressors
1327. Compressor valves are usually:
   a) thin, steel discs
   b) thick, steel discs
   c) a bi-metal material
   d) with handles on them
1328. Flywheels are usually used on:
   a) hermetic compressors
   b) shafts that are not tapered
   c) open type compressors
   d) rotary diaphragm compressors
1329. Rotary compressors have:
   a) pistons
   b) impellers
   c) horizontal pistons only
   d) either one or more blades
1330. Centrifugal compressors are used on:
   a) small residential units
   b) large air conditioning installations
   c) commercial sized equipment
   d) jobs requiring 50 or less tons
1331. Which temperature difference between the refrigerant and the surrounding air would be most desirable for a florist cabinet requiring high humidity?
   a) 7.5°F
   b) 15°F
   c) 20°F
   d) 25°F
1332. What's the special requirement for below-freezing temperature evaporators?
   a) must have two fans
   b) must be controlled by suction pressure

c) must have a defrost cycle
d) must be made of steel

1333. Where evaporators must be used to remove large amounts of humidity for air conditioning, which would be the most desirable coil from the standpoint of number of tubes deep (rows deep)?
    a) 2 tubes deep
    b) 4 tubes deep
    c) 6 tubes deep
    d) 8 tubes deep

1334. What is the name describing the plate evaporator where a solution for storage of cooling is provided?
    a) finned tube evaporator
    b) eutectic plates
    c) shell and tube
    d) receiver plate

1335. In a direct expansion chiller, is the refrigerant in the tubes or in the shell?
    a) tubes
    b) shell

1336. What type of metering device would be used on a flooded chiller?
    a) thermostatic expansion valve
    b) low-side float
    c) capillary tube
    d) hand expansion valve

1337. Which metering device will not shut off the flow of refrigerant when the compressor stops?
    a) thermal expansion valve
    b) high-side float
    c) low-side float
    d) capillary tube

1338. Which type of metering device has no moving parts?
    a) thermal expansion valve
    b) capillary tube
    c) high-side float
    d) low-side float

1339. Can the single model of thermostatic expansion valve be used for all refrigerants?
    a) no
    b) sometimes
    c) occasionally
    d) yes

1340. What type of relief device conserves refrigerant?
    a) spring loaded
    b) rupture disc

c) fusible plug
d) pressuretrol

1341. What type of marking should be used on a fusible plug?
    a) melting temperature
    b) pressure setting
    c) bursting pressure
    d) A.S. or E symbol

1342. Which is the highest when employing an oil safety switch?
    a) oil pressure
    b) back pressure

1343. Which of the following metering devices can not be used with a pressure control?
    a) expansion valve
    b) high-side float
    c) low-side float
    d) capillary tube

1344. What prevents leakage past the stem of a packless valve?
    a) a cap
    b) oil
    c) packing
    d) a diaphragm

1345. What prevents leakage through the packing of a low pressure valve?
    a) a cap
    b) oil
    c) packing
    d) diaphragm

1346. What type of automatic closing valves are used on a receiver sight glass?
    a) diaphragm type
    b) automatic ball-check valves
    c) packing tube
    d) solenoid type

1347. How large should the receiver be?
    a) tons of refrigeration times 5
    b) equal to volume of the system
    c) to the full of pump-down
    d) the full charge of the system should not occupy more than 85% of its volume

1348. Does an ammonia compressor use a plate or a screen baffle in the oil separator?
    a) plate
    b) screen

1349. With which refrigerant would water possibly be of assistance in the case of a leak?
    a) R-22

b) sulphur dioxide
c) ammonia
d) R-12

1350. What is the first step an operator should take in the event of a major fire involving refrigeration equipment?
   a) put the fire out
   b) call building manager
   c) evacuate the building
   d) shut down the equipment and call the fire department

1351. Under what circumstance would the Halide refrigerants become extremely hazardous when exposed to fire or flame?
   a) phosgene gas is produced
   b) refrigerant is light and rises rapidly
   c) refrigerant R-12 burns with a blue flame
   d) refrigerant R-22 will explode

1352. With which refrigerants is a gas mask required:
   a) Group I
   b) Group II
   c) Group III
   d) all groups

1353. Under what conditions are two or more masks required?
   a) Group I exceeds 500 lbs
   b) Group III exceeds 1000 lbs
   c) Group II exceeds 100 lbs
   d) any refrigerant exceeding 2000 lbs

1354. Where must the masks be stored?
   a) near compressors
   b) cabinet outside machinery room
   c) near refrigerant storage
   d) in a safe place

1355. How long can a third class operator be absent from his post of duty in any one hour to inspect the equipment related to the system?
   a) 5 minutes
   b) 10 minutes
   c) 15 minutes
   d) 20 minutes

1356. How much heat is required to raise 1 lb of ice from 0° to 32°F?
   a) 32 Btu
   b) 16 Btu
   c) 144 Btu
   d) 970 Btu

1357. How much heat is required to raise 1 lb of ice at 0° to steam at 212°F?
   a) 16 Btu
   b) 160 Btu

c) 340 Btu
d) 1310 Btu

1358. Additional heat added to steam above 212°F, at atmospheric pressure, is called:
    a) superheat
    b) latent heat
    c) heat of fusion
    d) heat of vaporization

1359. Which of the following is not a form of energy:
    a) chemical action
    b) heat
    c) electricity
    d) condensers

1360. Zero gauge pressure is equal to what absolute pressure?
    a) 30 inches vacuum
    b) 14.7 psia
    c) 30 psig
    d) 44.7 psia

1361. Twenty inches of vacuum is equivalent to what absolute pressure?
    a) 30.0 psig
    b) 59.7 psia
    c) -9.7 psig
    d) 5.0 psia

1362. Heat is not transmitted in which manner?
    a) insulation
    b) convection
    c) radiation
    d) conduction

1363. If a piston weighing 100 lbs exerts a force on an area of 20 in^2, what is the pressure exerted?
    a) 100 lbs
    b) 5 lbs/in^2
    c) 20 lbs
    d) 2 lbs/in^2

1364. Refrigeration:
    a) transfers heat
    b) produces cold

1365. Heat cannot be converted into:
    a) light
    b) water
    c) electricity
    d) work

1366. If a refrigerant is condensed at 194 lbs and 135°F, the liquid is sub-cooled at a temperature of:
   a) 140°F
   b) 120°F
   c) 180°F
   d) 160°F
1367. One atmosphere is equivalent to how many psia?
   a) 14.7
   b) 44.7
   c) 74.7
   d) 104.7
1368. If heat is added to a gas contained in a cylinder, the increase in pressure will be proportional to the increase in:
   a) temperature Fahrenheit
   b) temperature above 100°
   c) temperature Celsius
   d) absolute temperature
1369. In which part is there only liquid?
   a) suction line
   b) liquid line
   c) discharge line
   d) receiver
1370. How does an increase in evaporator pressure affect the boiling point of the refrigerant?
   a) no change
   b) raises it
   c) lowers it
1371. Which part is entirely on the high-side?
   a) compressor
   b) condenser
   c) metering device
   d) evaporator
1372. How does an increase in temperature in the liquid line affect the boiling pressure of the refrigerant?
   a) no change
   b) raises it
   c) lowers it
1373. In which components is the most latent heat dissipated?
   a) compressor
   b) condenser
   c) metering device
   d) evaporator

1374. What is the maximum amount of refrigerant that can be stored in a machinery room in addition to that stored in the system (not to exceed 300 lbs)?
   a) 20% of normal charge
   b) 30% of normal charge
   c) 40% of normal charge
   d) 50% of normal charge
1375. Are curves or tables best for finding the properties of refrigerant vapors?
   a) curves
   b) tables
1376. What equipment is used to test for a carbon dioxide refrigerant leak?
   a) halide torch
   b) lighted candle
   c) electronic detector
   d) R-12 gun
1377. How fast is the piston traveling in a modern high speed compressor?
   a) 150 rpm
   b) 300 rpm
   c) 450 rpm
   d) 500 rpm
1378. What does VSA refer to in compressor design?
   a) valves steel or aluminum
   b) vertical single-acting
   c) very slow acceleration
   d) vefur sataci atture
1379. Which is a method of capacity modulation?
   a) change in refrigerant charge
   b) valve adjustment
   c) two-staging
   d) cylinder loading
1380. An application of variable speed capacity controls would be:
   a) a low-temperature rotary compressor
   b) aircraft air conditioning
   c) automotive air conditioning
   d) a centrifugal compressor
1381. The disadvantage of strong compressor valve springs is:
   a) sluggish operation
   b) noise
   c) hard to lubricate
   d) short life
1382. Which pressures have the highest volumetric efficiency?
   a) high back pressures
   b) low back pressures
1383. The lowest capacity at which a centrifugal compressor can safely operate is what percent of full capacity?

- a) 10%
- b) 25%
- c) 35%
- d) 50%

1384. Why use safety heads on a compressor?
   - a) to assist valve operation
   - b) to increase the compression ratio
   - c) to prevent damage due to liquid slugging
   - d) to prevent piston slapping

1385. The capacity of an evaporative condenser is based on what air condition?
   - a) dry bulb temperature
   - b) wet bulb temperature
   - c) dew point temperature
   - d) relative humidity

1386. Ambient temperature below what amount must require some sort of head pressure control?
   - a) 60°F
   - b) 50°F
   - c) 40°F
   - d) 30°F

1387. During summer operation, cooling tower water is made available to the condenser at what temperature?
   - a) 75°F
   - b) 80°F
   - c) 85°F
   - d) 90°F

1388. What type of condenser cannot be cleaned with a brush or router?
   - a) double pipe
   - b) shell and tube
   - c) shell and coil

1389. Eliminator plates are used on:
   - a) cooling towers
   - b) evaporative condensers

1390. The most efficient type double pipe condenser is designed for:
   - a) parallel flow
   - b) counter flow

1391. Most water-cooled condensers, using cooling tower water as applied to air conditioning installation at design condition, raise the temperature of the water how many degrees?
   - a) 5°F
   - b) 10°F
   - c) 15°F
   - d) 20°F

1392. The water level is maintained in the sump of a cooling tower by a:
   a) by-pass valve
   b) hand valve
   c) solenoid valve
   d) float valve
1393. A bleed-off is used on which of the following equipment:
   a) air-cooled condenser
   b) water-cooled condenser
   c) evaporative condenser
   d) cooling tower
1394. Which type of metering device controls suction pressure in the evaporator?
   a) low-side float
   b) thermal expansion valve
   c) automatic expansion valve
   d) capillary tube
1395. What is the principal advantage of the hand expansion valve?
   a) it is load oriented
   b) it has simple construction
   c) it controls suction pressure
   d) it uses less refrigerant
1396. The thermostatic expansion valve operation is determined by how many pressures?
   a) 1
   b) 2
   c) 3
   d) 4
1397. On a thermal expansion valve, if the inlet pressure is 27 psig and the spring pressure is 7 psig, what should the bulb pressure be for the valve to be in equilibrium (valve has no equalizer connection)?
   a) 27 psig
   b) 7 psig
   c) 34 psig
   d) 20 psig
1398. A capillary tube must be used on what kind of a system?
   a) critically charged system
   b) flooded system
   c) receiver system
   d) multiple evaporator system
1399. The low-side float is used on what type of system?
   a) critically charged system
   b) flooded system
   c) pump-down system
   d) direct expansion system

1400. The low-side float has what function?
    a) control the superheat
    b) control the suction pressure
    c) minimize the use of refrigerant
    d) maintain the liquid level
1401. What should be the fin spacing on an evaporator used for a freezer room?
    a) 4 fins/inch
    b) 8 fins/inch
    c) 12 fins/inch
    d) 20 fins/inch
1402. The Celsius temperature equivalent to -40°F is:
    a) 0°C
    b) -20°C
    c) -40°C
    d) -60°C
1403. Temperature is:
    a) an indication of relative heat or cold
    b) the amount of heat to raise 1 lb of water 1 degree
    c) a quantity that causes an increase in temperature
    d) a feeling of warmth
1404. How does heat travel?
    a) warm to cold
    b) cold to warm
1405. How many Btu are necessary to raise 1 gallon of milk from 45°F to 80°F? (sp heat mile .92) (1 gal. milk weighs 8.6 lb)
    a) 125 Btu
    b) 177 Btu
    c) 225 Btu
    d) 277 Btu
1406. How many Btu are necessary to raise the temperature of 4 milk cartons from 45°F to 80°F? (sh. of milk carton = .32; one milk carton weighs .75)
    a) 4.8 Btu
    b) 33.6 Btu
    c) 8.4 Btu
    d) 36.3 Btu
1407. How many degrees Celsius is 75°F?
    a) 10°C
    b) 20°C
    c) 24°C
    d) 32°C
1408. Water will boil at 32°F if the surrounding pressure is:
    a) .089 psia
    b) .015 psia

c) .0019 psia
d) .0001 psia

1409. Does the heat loss from the condenser equal the heat gain to the evaporator?
    a) it is more
    b) it is less
    c) they are equal
    d) they are equal only after running for a certain time
    e) the heat gain is five times the heat loss

1410. Why is the condenser always purged after charging, if possible?
    a) to remove excess refrigerant
    b) to help the unit pump-down
    c) to remove non-condensable pressure
    d) it is not necessary to purge it
    e) to check the oil level

1411. What temperature should the head pressure of a water-cooled unit correspond to when the unit is running?
    a) 50° to 70°F above outlet water temperature
    b) 25° to 30°F above outlet water temperature
    c) 15° to 20°F above outlet water temperature
    d) 15°F above room temperature
    e) 25°F above room temperature

1412. What type of compressor do large water chillers usually have?
    a) centrifugal
    b) hermetic
    c) rotary
    d) reciprocating
    e) free piston

1413. What must one do to purge the condenser of a fuel gas operated water chiller?
    a) attach gauges
    b) isolate the condenser
    c) they have automatic purging devices
    d) crack the line at the condenser header
    e) it is not necessary to purge

1414. What is wrong if the low side pressure cannot be brought down to its normal setting?
    a) there is too much refrigerant in the system
    b) there is water in the system
    c) the TEV is too small
    d) there is dirt in the system
    e) the compressor is faulty

1415. Where is the liquid receiver usually located in a TEV system?
    a) there is none
    b) in the evaporator
    c) at the outlet of the condenser

d) in the suction line
e) in the filter

1416. What is wrong if the unit runs all the time and the evaporator is warm?
    a) bad relay
    b) bad thermostat
    c) bad connection
    d) too much refrigerant
    e) broken compressor valve

1417. Which two of the motor terminals will have the largest voltage drop or lowest amperage flow?
    a) starting to common
    b) starting to running
    c) running to common
    d) they are all the same
    e) two circuits are equal

1418. How is a motor winding open circuit detected?
    a) fuse will blow
    b) circuit breaker will open
    c) the unit will indicate a short
    d) neither winding will allow current to pass
    e) one winding will not allow current to pass

1419. Why must the electrical power source be checked?
    a) the wrong type wall socket may be used
    b) the outlet may be too far away
    c) the wiring may be overloaded
    d) the black and white wires may be interchanged

1420. Does the self-contained type of comfort cooler dehumidify the air?
    a) only if a humidistat is used
    b) only if the unit has filters
    c) only if the air is very warm
    d) yes
    e) no

1421. What does the overload motor cut-out protect?
    a) the thermostat
    b) the motor
    c) the wiring
    d) the compressor
    e) the filter

1422. What is one main advantage of a pressure-operated water valve?
    a) no cost of operation
    b) turns the water on and off
    c) easiest to install
    d) varies the rate of water flow
    e) will operate satisfactorily under high water pressure

1423. Where is the refrigerant line to the water valve usually connected into the system?
    a) suction line
    b) compressor crankcase
    c) receiver
    d) compressor head
    e) condenser
1424. What is the usual temperature difference between the water outlet and the water inlet?
    a) 0°F
    b) 10°F
    c) 30°F
    d) 40°F
    e) 70°F
1425. What happens to the water flow when the condensing unit stops?
    a) keeps on running
    b) continues running but at a lower rate of flow
    c) stops
    d) increases
1426. What is one of the advantages of an electric water valve?
    a) easy to install
    b) turns the water on and off
    c) no cost of operation
    d) can operate under high pressure
    e) varies the rate of water flow
1427. How do ice-makers produce clear ice cubes?
    a) distilled water is used
    b) the water is boiled first
    c) flowing water is frozen
    d) the water is fast-frozen
    e) the water is cooled to -10°F
1428. How is ice cube making stopped automatically?
    a) a bin-operated control is used
    b) the system produces a certain number of ice cubes and then stops
    c) the system stops when the water is used up
    d) a timer is used
    e) a pressure control is used
1429. What type of ice is produced when an auger is used?
    a) large sheets of ice
    b) solid cubes
    c) round, hollow cubes
    d) ice chips
    e) cone-shaped cubes

1430. What material is the evaporator usually made of on ice makers?
   a) aluminum
   b) copper
   c) plastic
   d) brass
   e) stainless steel

1431. What type of defrost system is used when solid, cubical ice cubes are made?
   a) hot gas
   b) electric
   c) thermal liquid
   d) electric and hot gas
   e) electric and thermal liquid

1432. How may one detect if too much refrigerant is in the system?
   a) take some out
   b) the low side pressure will be excessive
   c) the head pressure will be above normal
   d) liquid line will be warm
   e) the suction line will be warm

1433. Where is the two-temperature valve located?
   a) in the suction line
   b) in the liquid line
   c) in the condenser line
   d) in place of the TEV
   e) at the inlet to the evaporator

1434. What is wrong if the warmer evaporator is frosting back on a two-temperature system?
   a) the two-temperature valve is leaking
   b) the two-temperature valve is set too low
   c) the system is overcharged
   d) the TEV is leaking
   e) the thermostat has "stuck closed" points

1435. How is the warmer evaporator pressure checked?
   a) a gauge opening on the two-temperature valve
   b) it cannot be checked
   c) by using the gauge manifold
   d) by checking the suction temperature
   e) by removing, adjusting and then installing it

1436. Which evaporator has the greatest capacity?
   a) the warm evaporator
   b) the colder evaporator
   c) the two capacities are the same
   d) it does not make any difference
   e) only the total capacity is important

1437. How far should external and internal pipe threads overlap or contact when correctly installed on evaporative condensers?
   a) 3 threads
   b) 5 threads
   c) 7 threads
   d) 9 threads
   e) 11 threads

1438. What is meant by calibrating a float?
   a) adjusting to provide the proper water level
   b) heating in order to test for leaks in the float
   c) dehydrating the float
   d) measuring the float size
   e) weighing the float

1439. Why must water be added when using an evaporative condenser?
   a) it is cooler
   b) some is evaporated
   c) it is cleaner
   d) it is circulated faster
   e) it is under higher pressure

1440. What is the static head of a cooling tower system?
   a) the height of the unit
   b) the length of the pipe to the condenser
   c) the length of the pipe from the condenser
   d) the vertical height of the piping
   e) the water pressure when the pump is running

1441. Why is a dehydrator installed in a new system?
   a) to remove the copper chips
   b) to remove any moisture that may be in the system
   c) to keep moisture in the system from freezing
   d) to remove the air
   e) to keep the oil in the compressor

1442. How should the suction line be installed?
   a) slanted upward toward the compressor
   b) horizontally
   c) have vertical loops
   d) have no loops
   e) installed horizontally and vertically only

1443. What is probably wrong if the liquid line frosts?
   a) the system is short of refrigerant
   b) a float needle is leaking
   c) the system is too cold
   d) the screens in the liquid receiver are partially clogged
   e) there is too much refrigerant in the line

1444. What partial pinching causes the greatest decrease in the capacity of the unit?
   a) the suction line
   b) the liquid line
   c) the condenser line
   d) the evaporator coil tube
   e) they are all equal
1445. What happens to oil as it becomes colder?
   a) it is thicker
   b) it is thinner
   c) it congeals
   d) it precipitates
   e) it loses its lubricating properties
1446. Why are the drier and sight glass put in the system last?
   a) they do not need evacuating
   b) a vacuum cannot be pulled if they are in the system
   c) a vacuum ruins the dessicant
   d) a vacuum ruins the moisture indicator chemical
   e) to keep the drier and indicator free of impurities
1447. How can a nearby surface be protected from being scorched during a joint brazing operation?
   a) use a low temperature flame
   b) bend the pipe away from the surface
   c) use a metal or asbestos plate shield
   d) use flared connections
   e) refinish scorched surface
1448. Where should a frozen foods cabinet be coldest?
   a) in the center of the cabinet
   b) at the top rear inner wall
   c) at the top center of the cabinet
   d) on the coil surface
   e) the temperature should be the same all over
1449. What is considered a good frozen foods temperature?
   a) 32°F
   b) 15°F
   c) 5°F
   d) -10°F
   e) -20°F
1450. What is considered a good fast-freezing temperature?
   a) 32°F
   b) 10°F
   c) 5°F
   d) -10°F
   e) -20°F

1451. How should a dry coil be mounted?
    a) in a level position
    b) in the center of the cabinet
    c) 3" from the top of the cabinet
    d) at a slant toward the suction line
    e) as low as possible
1452. Before starting an installed unit, what should be done to the thermostatic expansion valve adjustment?
    a) turned out
    b) turned in
    c) nothing
    d) turned to left
    e) turned to right
1453. What is the average temperature difference between the evaporator refrigerant and the cabinet air?
    a) 0°F
    b) 10°F
    c) 20°F
    d) 30°F
    e) 40°F
1454. Where should the thermostat be located on a reach-in cabinet?
    a) on the coil
    b) in back of the coil
    c) on the cabinet wall
    d) just outside the door
    e) on the condensing unit
1455. What is one of the advantages of a blower coil in a reach-in cabinet?
    a) it requires little space
    b) it does not dry up foods
    c) the air movement is slow
    d) it is cheaper to operate
    e) it is easier to install
1456. How fast does the air flow over the typical blower coil in a reach-in cabinet?
    a) 55 fpm
    b) 1000 fpm
    c) 1500 fpm
    d) 2000 fpm
    e) 2500 fpm
1457. What is used to make a sweet water bath?
    a) tap water
    b) a salt brine
    c) no brine
    d) an alcohol brine

1458. What is the lowest temperature at which the refrigerant may be operated in a direct water cooler?
   a) 40°F
   b) 35°F
   c) 30°F
   d) 32°F
   e) 28°F
1459. To what temperature should water for restaurants be cooled?
   a) 55°F
   b) 40°F
   c) 45°F
   d) 35°F
   e) 32°F
1460. How do water coolers increase their efficiency with a heat exchanger?
   a) use a water-cooled condenser
   b) cool the incoming water with a heat transfer to the drain water
   c) use the drain water to cool the condenser
   d) use the incoming water to cool the condenser
   e) run the liquid line through the cooled water
1461. How many Btu must be removed to cool twenty gallons of water from 80°F to 40°F?
   a) 1000 Btu
   b) 2000 Btu
   c) 5550 Btu
   d) 6672 Btu
   e) 7777 Btu
1462. With what basic function of air conditioning are duct velocities most concerned?
   a) air heating
   b) air cooling
   c) air humidification
   d) air distribution
   e) air cleanliness
1463. What is meant by balancing a duct system?
   a) make all the openings equal in area
   b) make all the ducts equal in length
   c) make all openings discharge the correct amount of air
   d) make the inlet air equal to the outlet air
   e) adjust the duct dampers until all the air velocities are the same
1464. If the condensing unit is above the evaporator, why is a "U" bend put in the suction line?
   a) to prevent frost back
   b) to assist the oil return
   c) to prevent back pressure in the evaporator

d) to prevent high/low side pressures
e) to act as a vibration control
1465. What is the best way to clean a condenser in the home?
a) carbon tetrachloride
b) kerosene
c) brush
d) vacuum cleaner
e) cloth
1466. What is the best indication of a stuck-closed TEV?
a) no cooling
b) too cold
c) a partially frosted evaporator coil
d) a sweating or frosted suction line
e) a hot liquid line
1467. How may one find out if the thermostat is faulty when the unit won't start?
a) short the thermostat terminals
b) connect the power directly to the motor
c) a bad thermostat always causes continuous running
d) heat the thermostat bulb
e) cool the thermostat bulb
1468. What is wrong if the unit has an excessive head pressure with a normal low side pressure?
a) too much refrigerant
b) stuck-closed TEV
c) air in the system
d) room too cold
e) wrong refrigerant
1469. Where is the power bulb located when controlling an air-cooled evaporator?
a) on the coil
b) on the liquid line
c) on the suction line near the coil
d) on the suction line outside the cabinet
e) anywhere on the cabinet
1470. What does the thermostatic expansion valve control?
a) the temperature of refrigerant in the coil
b) the quantity of refrigerant in the coil
c) the pressure in the coil
d) the temperature in the coil
e) the high side pressure
1471. What will happen to the coil if the power element is located in a warm air stream?
a) flood
b) starve
c) nothing

d) get too cold
e) get too warm
1472. What temperature should the head pressure correspond to when the unit is running?
    a) 70°F
    b) 10°F above the room temperature
    c) 35°F below the room temperature
    d) 35°F above the room temperature
    e) it is independent of the temperature
1473. What problem is indicated when the liquid line is warmer than the room temperature?
    a) an overcharge of refrigerant
    b) a stuck-open TEV
    c) a lack of refrigerant
    d) too much oil
    e) a low-side pressure that is too high
1474. Is there any oil in the condenser?
    a) yes
    b) only if the system is water-cooled
    c) only if too much oil is in the system
    d) no
    e) only when the unit is being pumped down
1475. How is a normal head pressure maintained in sub-zero weather?
    a) by an undersized condenser
    b) by overcharging the system
    c) by restricting the air flow over the condenser
    d) no change has to be made
    e) by shutting off the condenser fan
1476. What is the correct reasoning when there is a voltage drop at a terminal?
    a) there is a loose, dirty connection
    b) the terminal is too large
    c) the electrical load is too small
    d) the voltmeter is connected incorrectly
    e) the terminal is grounded
1477. Why must the wiring be checked?
    a) it must be strong enough
    b) it must be able to carry the voltage imposed
    c) it must be able to carry the current
    d) it must be able to carry the running current
    e) it should not be too large
1478. How should a voltmeter be connected into a circuit?
    a) in series
    b) in parallel
    c) its prongs are clamped around the insulated wire

d) it is used only at the power source
e) it is used with a shunt

1479. What material is usually used for compressor cylinders?
- a) aluminum
- b) copper
- c) cast iron
- d) steel
- e) brass

1480. How many inches of vacuum should a good compressor be capable of creating?
- a) 36"
- b) 28"
- c) 24"
- d) 15"
- e) 10"

1481. Which type of piston crankshaft arrangement does not use a connecting rod?
- a) Scotch yoke
- b) eccentric
- c) conventional
- d) hotchkiss
- e) "V" type

1482. Which type of compressor has the cylinders parallel to the crankshaft?
- a) swash plate
- b) eccentric
- c) conventional
- d) Scotch yoke
- e) hotchkiss

1483. Where are compressor valves usually located on open-type compressors?
- a) in the cylinder head
- b) in a valve plate
- c) in the piston head
- d) in the tubing
- e) no valves are needed

1484. What is the most difficult internal electrical trouble to detect in motors?
- a) a short
- b) a ground
- c) a burned winding
- d) a broken line
- e) a lack of continuity

1485. What size fan motor should be used as a replacement?
- a) 1/10 hp more
- b) 1/10 hp less
- c) one size smaller
- d) the same size

e) one size larger
1486. How can air be cleaned directly using electricity?
   a) heat the air
   b) magnetize the air
   c) cool the air
   d) electrical ionization
   e) by high speed air movement
1487. Do all filters have a pressure drop?
   a) yes
   b) no
   c) only if they are dirty
   d) only if high air speeds are used
   e) only the water spray type
1488. Do all the particles of air travel at the same speed in the duct?
   a) no
   b) yes
   c) only if they are the same size
   d) only in round ducts
   e) only if the velocity is low
1489. Are the pressures the same throughout the duct systems?
   a) no
   b) yes
   c) only if a slow speed fan is used
   d) the pressures are higher at the inlet
   e) only if there is no turbulence
1490. What effect does a damper have on air flow?
   a) improves it
   b) retards it
   c) streamlines the air flow
   d) maintains equal pressures
   e) prevents overloading the fan
1491. If all the outlet openings are the same size, what is the result?
   a) each opening will discharge an equal volume of air
   b) each outlet will discharge an equal weight of air
   c) the farthest opening will discharge the least volume of air
   d) the closest opening to the fan will discharge the least volume of air
   e) the closest opening to the fan will discharge all the air
1492. What is considered a maximum comfortable draft?
   a) 5 fpm
   b) 10 fpm
   c) 15 fpm
   d) 20 fpm
   e) 25 fpm

1493. What is another instrument used to obtain air velocity values?
- a) hydrometer
- b) speedometer
- c) pedometer
- d) odometer
- e) anemometer

1494. Why is the velocimeter popular for determining air velocities?
- a) it does not need the 16 positions
- b) it can be mounted in any position
- c) it is more accurate than the pitot tube
- d) it is excellent for very low velocities
- e) the velocities can be read directly

1495. Can the velocimeter be used to obtain velocities inside a duct?
- a) no
- b) yes
- c) yes, if the instrument can be put inside
- d) yes, if the duct is large enough for the person to get inside
- e) yes, but only if rough readings are desired

1496. Does the velocimeter operate on the principle of velocity pressure?
- a) yes
- b) no
- c) only at high velocities
- d) only at outlet grilles
- e) only at inlet grilles

1497. If a 12" x 12" grille has 40 metal strips 1/32" thick mounted lengthwise, what is its net opening?
- a) 144 in^2
- b) 156 in^2
- c) 129 in^2
- d) 141.75 in^2

1498. What produces the lower reading of the wet bulb?
- a) cooling by evaporization
- b) the thermometer is calibrated that way
- c) the cloth and water form a cooling solution
- d) it doesn't read lower
- e) ice

1499. Why are thermometers moved through the air?
- a) to remove excess liquid water
- b) to enable the thermometers to contact more air and then obtain a more average reading
- c) to prevent the vapor saturating the air immediately around the wick
- d) to deep the thermometer mercury from separating
- e) to obtain an average of temperature stratification

1500. What does wet bulb depression mean?
- a) a cavity in the bulb
- b) the wet bulb is located below the dry bulb
- c) the difference in readings between the wet bulb thermometer and the dry bulb thermometer
- d) the inaccuracy of the thermometer
- e) the appearance of mercury in the thermometer

SECTION TWO

HEATING

1501. The leakage of air around windows, doors and cracks in a house is:
   a) an air inlet
   b) infiltration
   c) a by-pass
   d) psia
1502. A limiting device installed on a forced air furnace is controlled by:
   a) a vent
   b) a by-pass
   c) temperature
   d) pressure
1503. One cubic foot of natural gas would be:
   a)   100 Btu
   b)   2,500 Btu
   c)   1,000 Btu
   d) 140,000 Btu
   e)   250 Btu
1504. How much air is needed to burn 1,000 Btu?
   a)   100 ft^3
   b)   10 ft^3
   c) 1,000 ft^3
   d)   250 ft^3
   e)   1 ft^3
1505. The water pressure (in inches) from the gas meter to the appliance would be:
   a) 4 to 5" wc
   b) 3 to 4" wc
   c) 8 to 11" wc
   d) 6 to 7" wc
   e) 9 to 11" wc
1506. The manifold pressure of natural gas would be:
   a) 3 to 4" wc
   b) 6 to 7" wc
   c) 8 to 11" wc

d) 4 to 5" wc
e) 9 to 11" wc

1507. What is the permissible pressure drop between the meter and the appliance?
   a) 3.0" wc
   b) 50.0" wc
   c) 5.0" wc
   d) 0.50"wc
   e) 11.0" wc

1508. An oil tank of 220 gallons must be what gauge?
   a) 14 ga
   b) 12 ga
   c) 16 ga
   d) 10 ga
   e) 8 ga

1509. On a low pressure residential forced-water boiler, the relief device would open up at:
   a) 15 lbs
   b) 30 lbs
   c) 125 lbs
   d) 150 lbs

1510. On a residential low pressure forced-water boiler, the pressure reducing valve would be factory set at:
   a) 30 lbs
   b) 15 lbs
   c) 12 lbs
   d) 125 lbs

1511. On a residential low pressure steam boiler, the pressure reducing valve would be set at:
   a) 30 lbs
   b) 15 lbs
   c) 12 lbs
   d) none of the above

1512. On residential low pressure forced-water boilers, you must have, on new installations, a gate valve at the:
   a) outlet of the boiler
   b) inlet of the boiler
   c) outlet and inlet of the boiler
   d) none of the above

1513. On a residential low pressure steam boiler, the device used for expansion of water would be:
   a) a stack pipe
   b) a compression tank
   c) an airtrol
   d) none of the above

1514. On a low pressure residential steam boiler, the relief device opens up at:
    a) 30 lbs
    b) 12 lbs
    c) 15 lbs
    d) none of the above
1515. The type of zone valve for White Rogers would be:
    a) diaphragm
    b) heat-activated
    c) motorized
    d) none of the above
1516. Minneapolis Honeywell zone valves are:
    a) motorized
    b) heat-activated
    c) diaphragm
    d) none of the above
1517. Normal operating pressure for residential forced-water boilers would be:
    a) 30 lbs
    b) 12 to 15 lbs
    c) 15 to 30 lbs
    d) none of the above
1518. The rate at which heat flows through the composite of several different materials used in a building is called the "U" or:
    a) temperature factor
    b) maintained temperature
    c) heat transmission factor
1519. Why does a pilot light have a safety device?
    a) to relight the pilot light
    b) to stop the pilot light gas flow if there is no pilot flame
    c) to stop the main gas flow if there is no pilot flame
    d) to shut off the furnace if the stack temperature is too high
    e) to shut off the furnace if the furnace temperature is too high
1520. Where is the primary air adjustment?
    a) at the pilot light
    b) at the gas spud
    c) at the entrance to the burner
    d) at the pressure regulator outlet
    e) at the entrance to the gas manifold
1521. What circuits does the relay operate in an intermittent ignition primary control?
    a) the motor and the stack bimetal circuits
    b) the ignition and the thermostat circuits
    c) the ignition and the motor circuits
    d) the ignition transformer
    e) the overload manual reset circuits

1522. Where is the step-down transformer located?
   a) in the limit switch
   b) in the thermostat
   c) in the primary control
   d) in the ignition transformer
   e) in the manual switch box
1523. The average gravity warm-air heating system needs:
   a) a return for each room
   b) usually two or three returns
   c) no returns
   d) five or more returns
1524. The most difficult style of house to heat is probably the:
   a) two story
   b) ranch style
   c) split level
   d) six-room single level
1525. The gas supply connection from the pilot shall be made:
   a) between the main shut-off valve and the meter
   b) between the main shut-off valve and the burner
   c) by drilling and tapping the gas supply line
   d) at any convenient location
1526. The secondary air adjustment shutter should always be:
   a) locked in the wide open position
   b) properly adjusted by the installer
   c) left in the position set by the factory
   d) replaced by a barometric damper
1527. Delayed gas ignition may be caused by:
   a) a defective thermostat
   b) an unvented pressure regulator
   c) a blown fuse
   d) an improperly located pilot
1528. A separate electrical permit is required:
   a) for all gas heating installations
   b) for any connections to the 110 volt circuit other than the installation of a transformer on an existing outlet
   c) when an electrician wires the installation
   d) when a transformer is installed
1529. How can a low voltage thermostat be used in an electrical resistance heating system?
   a) use a low voltage power circuit
   b) use a low amperage circuit
   c) use a relay
   d) use parallel circuits
   e) use series circuits

1530. How is a tubular electrical resistance heating unit insulated?
   a) it is not
   b) plastic insulation is used
   c) the resistance wire is overlaid in an oxide powder
   d) ceramic insulators are used
   e) porcelain insulators are used

1531. How many grains of moisture will saturate one pound of dry air at 0°F?
   a) 5.5 grains
   b) 10.5 grains
   c) 15.5 grains
   d) 20.5 grains
   e) 25.5 grains

1532. How many grains of moisture will saturate one pound of dry air 50% at 75°F?
   a) 5.5 grains
   b) 36 grains
   c) 50 grains
   d) 66 grains
   e) 132 grains

1533. Which control devices may be used on the 24 volt circuit in a domestic gas-fueled heating hydronic system?
   a) thermostat
   b) thermostat, relay and safety pilot
   c) thermostat, relay and gas solenoid
   d) thermostat, relay, gas solenoid, limit control and safety pilot
   e) thermostat, relay, gas solenoid, limit control, safety pilot and circulation pump

1534. The factor or factors used to determine the percentage of the total heat loss of a home by outside walls and windows is:
   a) 60% to 80%
   b) 100%
   c) 50%
   d) 25% to 50%

1535. The flue pipe from a gas-designed furnace shall be:
   a) at least 8" in diameter
   b) the size indicated by the flue connection on the furnace
   c) one size larger than the flue connection on the furnace
   d) any convenient size

1536. The flue pipe from a gas-fired hot water heater should:
   a) be connected to the gas furnace flue pipe ahead of the diverter
   b) be connected to the gas burner flue pipe behind the diverter
   c) enter the chimney through a separate opening
   d) be removed from the chimney

1537. In case of a ruptured diaphragm in a pressure regulator, the escaping gas should:
   a) discharge into the furnace room
   b) cause the solenoid valve to close
   c) be ignited by the pilot flame
   d) cause an explosion

1538. On gas equipment of 400,000 Btu or more, if the gas pilot flame goes out, the safety pilot should shut off the gas to the main burner within:
   a) 2 minutes
   b) 3 minutes
   c) 10 seconds
   d) 60 seconds

1539. What may happen to a flame if too much primary air is used?
   a) the flame will be too short
   b) the flame will lift off the burner
   c) the flame will be too hot
   d) the flame will be irregular

1540. Supply outlets (diffusers or registers) introduce the conditioned air into the:
   a) rooms
   b) furnace plenum
   c) wall stacks
   d) attic

1541. An area in which fuel or gaseous substance is burned is called a:
   a) combustion chamber
   b) firebox liner
   c) bonnet
   d) radiator

1542. An accurate way to check the input of a gas burner on a residential forced-water boiler would be:
   a) reading the Btu input of the boiler
   b) reading the output of the Btu on the boiler
   c) reading the net Btu of the boiler
   d) clocking the revolutions of the gas meter dial

1543. The temperature of steam at 5 psig pressure would be:
   a) 212°F
   b) 180°F
   c) 230°F
   d) 205°F

1544. The purpose of the deflector plate in a draft diverter is to:
   a) provide for adjustment of pressure level
   b) prevent a back draft from entering the combustion chamber
   c) deflect the flue gases

1545. The setting of the limit control on a hot water supply heater or boiler must not exceed:
   a) 200°F
   b) 250°F
   c) 210°F
   d) 212°F
1546. Room heating units, baseboard radiation, convectors, radiators, and etc. are sized to the heat loss of:
   a) the room
   b) the boiler
   c) inside rooms
   d) none of the above
1547. A steam trap is a device which will allow steam, but not condensation, to pass through it.
   a) true
   b) false
1548. Soot deposits in a boiler would cause temperatures in a stack to:
   a) remain the same
   b) increase
   c) decrease
1549. The term convector, as used in hydronics, refers to a unit which emits the greater portion of its heat by:
   a) radiation
   b) conduction
   c) convection
   d) none of the above
1550. If the gas pilot goes out for any reason, the safety pilot on equipment of less than 400,000 Btu, the gas to the main burner should shut within:
   a) 10 seconds
   b) 90 seconds
   c) 3 minutes
   d) 2 minutes
1551. The steam temperature of 5 inches of Hg vacuum would be:
   a) 180°F
   b) 160°F
   c) 205°F
   d) 212°F
1552. How much heat does one pound of steam release as it is condenses to water at 212°F?
   a) 144 Btu
   b) 212 Btu
   c) 1000 Btu
   d) 970 Btu

1553. What happens inside the pipe of a single-pipe, residential, low-pressure steam system?
   a) water travels to the boiler
   b) steam travels to the radiator
   c) steam travels to the radiator while water drains back to the boiler
   d) steam travels during heating, water returns during off cycle

1554. Oil tank vent pipes on tanks of 275 gallons must be at least:
   a) 1"
   b) 1 ¼'
   c) 1 ½'
   d) 2"
   e) 2 ½'

1555. Oil tank fill pipes on tanks of 275 gallons must be at least:
   a) 1 ft above grade
   b) 2 ft above grade
   c) 3 ft above grade
   d) 4 ft above grade
   e) 5 ft above grade

1556. Oil tank vent pipes should terminate outside the building and be at least:
   a) 1 ft above grade
   b) 2½ ft above grade
   c) 3 ft above grade
   d) 3½ ft above grade
   e) 4 ft above grade

1557. On most circulators the number of oil cups would be:
   a) one
   b) two
   c) three
   d) four

1558. On a B & G circulator, the term B & G stands for:
   a) ball and gossett
   b) bell and gossett
   c) ball and green
   d) bell and green

1559. On residential circulating pumps, the device that moves the water is called:
   a) a coupler
   b) a bearing assembly
   c) an impeller
   d) a flange

1560. On a residential circulating pump, the device that is connected between the motor and the bearing assembly would be:
   a) a mechanical seal
   b) a bronze bearing

c) a coupler
d) an impeller

1561. During normal operation, flow control valves should be kept in the:
   a) open position
   b) closed position
   c) half-way open position

1562. B & G zone control valves are:
   a) motorized
   b) heat-activated
   c) none of the above
   d) all of the above

1563. The system which provides a channel for air passage to and from a furnace is called the:
   a) pipeline
   b) duct system
   c) passageway
   d) tunnel

1564. Heating system trunk ducts carry the air which is to be distributed to more than one:
   a) supply inlet
   b) furnace
   c) residence
   d) supply outlet

1565. If part of a supply duct of a domestic heating system passes through an unexcavated space or a masonry wall, that portion of it should be covered with:
   a) duct tape
   b) insulation
   c) vinyl film
   d) aluminum foil

1566. To combat added resistance of air flow (such as additional duct work) in a heating system, the most advisable remedy is to:
   a) speed up a small blower motor
   b) slow down the blower motor
   c) install a smaller blower assembly
   d) install a larger blower assembly

1567. What would you expect to find inside an extrol tank?
   a) an air pocket
   b) a hard rubber ball
   c) a vacuum
   d) a rubber diaphragm full of air

1568. The installation of a gas burner in a hot-water boiler must include:
   a) a low water cut-off
   b) a closed expansion tank

    c) a water temperature limit control
    d) an automatic water feeder

1569. The letter "H" on oil nozzles means:
    a) hollow
    b) solid
    c) special design
    d) semi-hollow

1570. What is the color of a correctly adjusted natural gas flame?
    a) orange
    b) red
    c) yellow
    d) blue
    e) green

1571. What happens if the air flow through a baseboard electric resistance heater is stopped?
    a) the thermostat will open the circuit
    b) the relay will open the circuit
    c) the system will continue to operate
    d) a timer will shut off the unit
    e) a limit control will open the circuit

1572. How much heat is created by one watt?
    a) 760 Btu
    b) 3,415 Btu
    c) 3.42 Btu
    d) 2545.6 Btu
    e) 14.4 Btu

1573. What is the recommended maximum ampere flow in an electrical resistance heating circuit?
    a) 15 amperes
    b) 20 amperes
    c) 25 amperes
    d) 30 amperes
    e) 50 amperes

1574. What is the relative humidity at 75°F db and 68°F wb effective temperature?
    a) 20%
    b) 30%
    c) 40%
    d) 50%
    e) 60%

1575. Why do some power humidifiers have electric motors?
    a) to operate the valve
    b) to stir the water
    c) to expose more water surface to the air

d) to pump water into the humidifier
e) to scrape the scale out of the humidifier pan

1576. What type of material varies in size as the humidity changes?
   a) ceramic
   b) metallic
   c) hydroscopic
   d) silicon
   e) hydraulic

1577. Ducts located in attics, ventilated crawl spaces or other spaces exposed to outdoor weather conditions should be:
   a) coated with paint to prevent corrosion
   b) coated with asphalt paint to reduce heat loss
   c) wrapped with at least three inches of insulation
   d) wrapped with at least two inches of insulation

1578. Return air grilles for perimeter heating systems may be located in the:
   a) floor and baseboard
   b) ceiling
   c) high in the inside walls
   d) all of the above

1579. What is the most common use of a clock in a heating system?
   a) to change from heating to cooling
   b) to turn the heating system on
   c) to reduce operating temperatures during the night
   d) to turn the cooling system on
   e) to operate the sequency controls

1580. A domestic heating load must be determined on a:
   a) room-by-room basis
   b) wall-to-wall basis
   c) floor-to-floor basis
   d) floor-to-ceiling basis

1581. What is the normal wattage capacity of a 120V electric heating thermostat?
   a) 15 watts
   b) 20 watts
   c) 746 watts
   d) 2000 watts
   e) 5000 watts

1582. What voltage is used on the limit control of a gun type oil burner system?
   a) 120 volts
   b) 24 volts
   c) 15 volts
   d) either 24 volts or 120 volts
   e) a 24 volt solenoid operates a 120 volt system

1583. How can a low voltage thermostat operate a 120 volt system?
   a) use a solenoid valve
   b) use a transformer
   c) use a relay
   d) use a sequency control
   e) use a resistor
1584. How is a 24 volt circuit supplied?
   a) from the power company
   b) use a solenoid
   c) use a thermostat
   d) use a step-up transformer
   e) use a step-down transformer
1585. In a low voltage system, what voltage operates the heat anticipator?
   a) 24 volts
   b) 6 volts
   c) 120 volts
   d) 12 volts
   e) 240 volts
1586. Where is a limit control located in the electrical circuits?
   a) in the 24 volt circuit
   b) in the 120 volt circuit
   c) in the 240 volt circuit
   d) in both the 24 volt and the 120 volt circuits
   e) in either the 24 volt or the 120 volt circuit
1587. How is velocity pressure measured with a pitot tube?
   a) add the static pressure and total pressure
   b) subtract the static pressure from the total pressure
   c) add the static pressure and atmospheric pressure
   d) subtract the atmospheric pressure from the static pressure
   e) subtract the total pressure from the static pressure
1588. What are the minimum number of pitot tube readings one should take in a rectangular duct?
   a) 4
   b) 8
   c) 10
   d) 12
   e) 16
1589. The pressure level in a gas conversion burner installed should be:
   a) below the fire door
   b) not over 5 lbs/in^2
   c) at approximately the center of the fire door
   d) limited by a pressuretrol
1590. Which instrument is used to measure outlet grille air velocities?
   a) anemometer

b) barometer
c) manometer
d) micrometer
e) draft gauge

1591. What is the average effective area of an outlet grille?
 a) 50%
 b) 60%
 c) 70%
 d) 80%
 e) 90%

1592. Why must an air duct elbow be considered as having an equivalent length?
 a) the elbow is smaller
 b) the air flow friction is lower
 c) the air flow friction is higher
 d) to know its curved length
 e) because all ducts have an equivalent length

1593. How should thermostat contact points be cleaned?
 a) sandpaper
 b) file
 c) paper
 d) stone
 e) emery cloth

1594. Which of the following series of burner controls indicates a three-wire 24 volt system?
 a) series 10
 b) series 20
 c) series 40
 d) series 80
 e) series 100

1595. To determine the entire distribution system layout and size, it is necessary to make a:
 a) blueprint survey
 b) heat loss survey
 c) structure estimate survey
 d) transmitting surface survey

1596. What is the sensible heat of dry air?
 a) 0.24 Btu/ft^3
 b) 0.24 Btu/lb
 c) 0.32 Btu/ft^3
 d) 1.43 Btu/lb
 e) 0.043 Btu/lb

1597. How much pressure drop is acceptable across a domestic filter?
 a) ½" water
 b) 1" water

c) 2" water
d) 3" water
e) 6" water

1598. What control is used to operate the fan motor?
   a) room thermostat
   b) primary control
   c) stack thermostat
   d) bonnet thermostat
   e) outdoor thermostat

1599. What is the return air volume compared to the warm air volume?
   a) less
   b) more
   c) same
   d) 10% more
   e) 20% more

1600. What should be the average main duct temperature of a warm-air system?
   a) 100°F
   b) 120°F
   c) 140°F
   d) 160°F
   e) 180°F

1601. The check opening in the fire door of a gas converted furnace should be:
   a) locked partly open
   b) left as is for adjustment
   c) locked fully open
   d) sealed closed

1602. Which of the following is not one of four significant factors affecting heat loss from a duct?
   a) temperature in the furnace bonnet
   b) rate of heat loss per foot of pipe
   c) type of fuel used in combustion to produce heat
   d) distance from the furnace to warm air registers
   e) how fast the air travels through the pipe

1603. How is hot water for domestic consumption often heated when a hydronic system is used?
   a) a separate hot water heater
   b) a heat exchanger
   c) the system water is used
   d) the pipe within a pipe is used
   e) the reserve water in the compression tank is used

1604. What is the most probable cause of oil around the base of the gun-type oil burner?
   a) improper combustion
   b) loose filter cover

c) excessive pressure
d) leaking pump shaft seal
e) dirty nozzle

1605. The gas supply line to the gas burner must be:
    a) not over 10 feet in length
    b) an independent line from the meter
    c) not over 30 feet in length
    d) installed by a plumber

1606. The proper method of checking for a suspected gas leak is:
    a) to depend upon the sense of smell
    b) with a match or small flame
    c) by sense of touch
    d) with soap and water

1607. What keeps the oil pressure constant at the nozzle?
    a) the oil level in the tank
    b) the pressure relief valve
    c) the oil pump
    d) the size of the nozzle orifice
    e) the pressure regulator

1608. What is the usual oil pressure at the nozzle in a high-pressure gun-type oil burner?
    a) 15 psig
    b) 30 psig
    c) 50 psig
    d) 100 psig
    e) 200 psig

1609. To what temperature must the atomized oil be heated to start combustion?
    a) 72°F
    b) 100°F
    c) 212°F
    d) 400°F
    e) 700°F

1610. Why is it important to keep water out of the oil burner oil passages?
    a) it causes rust
    b) it causes corrosion
    c) it is heavier than oil
    d) it will overload the motor
    e) it may cause a "flame out"

1611. Gas piping which is to be concealed must be:
    a) of extra, extra heavy pipe
    b) installed by a plumber
    c) provided with unions
    d) inspected before being concealed

1612. What is the efficiency of a gun-type oil burner installation if the stack temperature is 700°F and the percent of CO_2 is equal to 10%?
   a) 68.5%
   b) 71.5%
   c) 73%
   d) 74%
   e) 76.5%

1613. How much excess air is needed to produce 8% CO_2 in an oil burner system?
   a) 30%
   b) 50%
   c) 80%
   d) 125%
   e) 150%

1614. How is a smoke test made using the Ringlemann Scale?
   a) binoculars are used to observe stack fumes
   b) a sample is collected in distilled water
   c) a sample is forced through a white filter
   d) a filter is put in the stack for a definite time
   e) the soot deposit on stack internal wall is inspected

1615. How can the amount of excess air be determined?
   a) by measuring the amount of air fed to the furnace
   b) by using a draft gauge
   c) by using a pitot tube
   d) by using a stack CO_2 indicator
   e) by using a stack oxygen indicator

1616. A hydronic hot-water system has a compression tank to:
   a) create a pressure
   b) allow for expansion of the water
   c) fill the system when the water level is low
   d) collect the sediment in the system
   e) provide reserve heat

1617. What type of pump is used on a hydronic system?
   a) centrifugal
   b) piston
   c) rotary
   d) diaphragm
   e) gear

1618. In a low voltage system, what voltage operates the heat anticipator?
   a) 24 volts
   b) 6 volts
   c) 120 volts
   d) 12 volts
   e) 20.5 volts

1619. Basement heat loss is usually determined on the basis of measurements of the heat loss on:
   a) the wall above the grade or outside ground level
   b) the wall below the grade and the floor
   c) the basement floor
   d) items a and b
1620. A publication which contains information about heat transfer factors, heat transfer multipliers and "U" factors is:
   a) MHAW's Encyclopedia of Heat
   b) Webster's New Collegiate Dictionary
   c) Manual "J" of the NESCA
   d) the Heating Specialist's Guide
1621. The part of the casing which forms a mixing chamber from which supply ducts receive warmed air is called the:
   a) tight casing
   b) combustion chamber
   c) firebox
   d) bonnet
1622. The average temperature of the air discharged from the furnace outlet duct (or ducts) is called the:
   a) outlet air temperature
   b) inlet duct temperature
   c) stack loss
   d) duct temperature
1623. The average temperature of air entering the forced-air furnace is the:
   a) flue gas temperature
   b) inlet duct temperature
   c) bonnet efficiency
   d) fuel heat output
1624. The amount of heat input which escapes in flue gases is called:
   a) inlet duct temperature
   b) outlet air temperature
   c) stack loss
   d) chimney vapor
1625. To keep moisture from seeping into the heating system in houses built over crawl spaces, the ground should first be covered with:
   a) sawdust
   b) concrete
   c) vapor barrier
   d) gravel
1626. How many electrode installation dimensions should be checked?
   a) one
   b) two
   c) three

d) four
e) five

1627. On residential low-pressure steam boilers, normal operating pressure is:
   a) 30 lbs
   b) 12 to 15 lbs
   c) 15 to 30 lbs
   d) none of the above

1628. On Taco circulating pumps the bearing assembly is lubricated with:
   a) #20 non-detergent motor oil
   b) grease
   c) water
   d) none of the above

1629. Residential low-pressure forced-water boilers are sized:
   a) to the load of the building
   b) 10% greater than the load of the building
   c) 10% less than the load of the building

1630. On residential low-pressure forced-water boilers, the most important rating in dealing with load calculation is:
   a) Btu input
   b) Btu output
   c) Btu net
   d) none of the above

1631. On residential low-pressure steam boilers, low water cut-offs are required.
   a) true
   b) false

1632. On residential low-pressure forced-water boilers, low water cut-offs are not required.
   a) true
   b) false

1633. The amount of 3/4" copper pipe it would take to equal 1 gallon of water would be approximately:
   a) 13 ft
   b) 23 ft
   c) 33 ft
   d) 43 ft

1634. Residential low-pressure forced-water fin tubing (baseboard radiation) would be sized by:
   a) Btu input of the boiler
   b) Btu output of the boiler
   c) Btu net
   d) none of the above

1635. The maximum allowable temperature for residential low-pressure forced-water boilers is:
   a) 212°F

b) 210°F
c) 250°F
d) 200°F

1636. On residential low-pressure forced-water boilers, a type of pilot safety that could be found would be:
    a) hydraulic
    b) diaphragm
    c) motorized

1637. What effect does the fan have on the air?
    a) it cools it
    b) it heats it
    c) it cleans it
    d) it dries it
    e) nothing

1638. What type of air pressure is measured by a differential manometer?
    a) velocity pressure
    b) total pressure
    c) static pressure
    d) vacuum
    e) partial vacuum

1639. What is the pressure difference across an adhesive filter which indicates when the filter should be changed?
    a) 0.10" water
    b) 0.20" water
    c) 0.30" water
    d) 0.40" water
    e) 0.50" water

1640. If the filter is on the intake side of the fan, what is the pressure on both sides of the filter?
    a) positive pressure
    b) negative pressure
    c) both positive and negative pressure
    d) either positive or negative pressure
    e) negative pressure on one side of filter only

1641. When the air flow into a duct is reduced, what happens to the pressure drop across the filter?
    a) it stays the same
    b) it increases
    c) it decreases
    d) it may either increase or decrease
    e) it reduces to zero

1642. What does one inch of water column equal in psi?
    a) 29.9 psi
    b) 14.7 psi

c) 0.432 psi
d) 0.036 psi
e) 2.31 psi

1643. When a combination heating and cooling unit has an electronic air cleaner, it is recommended that the electronic air cleaner should operate:
   a) only in the heating cycle
   b) only when the indoor fan is operating
   c) only in the cooling cycle
   d) only by itself when neither heat or cooling is on
   e) none of the above

1644. In a forced air system, an electronic air clener would be installed:
   a) in the plenum chamber
   b) in the main trunk line
   c) on the return air side of the furnace fan
   d) at the return registers

1645. Wet bulb depression means:
   a) a cavity in the bulb
   b) the wet bulb is located below the dry bulb
   c) the difference in readings between the wet bulb thermometer and the dry bulb thermometer
   d) the inaccuracy of the thermometer
   e) the appearance of mercury in the thermometer

1646. Mean temperature indicates the:
   a) average temperature
   b) lowest temperature
   c) highest temperature
   d) design temperature
   e) dry bulb temperature

1647. Ambient temperature means:
   a) evaporator temperature
   b) condenser temperature
   c) surrounding temperature
   d) desired temperature
   e) dew point

1648. How many heat sources must be considered when calculating the cooling heat load?
   a) one
   b) two
   c) three
   d) four
   e) five

1649. The most efficient type of duct is:
   a) round
   b) square

c) rectangular
   d) oblong
   e) all the same
1650. A duct fitting which adapts the duct to a wall stack, register or grille is a:
   a) plenum
   b) duct sleeve
   c) boot
   d) take-off
   e) none of the above
1651. Why isn't an anemometer used for measuring drafts?
   a) it is too accurate
   b) it won't operate on slow air velocities
   c) it always indicates too high draft velocities
   d) an anemometer measures temperature
   e) anemometers only operate inside ducts
1652. When do air velocities out exceed the air velocities in?
   a) never
   b) always
   c) only if the inlet area is larger than the outlet area totals
   d) only when the inlet area is smaller than the outlet area totals
   e) only when a high velocity fan is used
1653. A fluctuating water level is usually caused by:
   a) too high a load
   b) foaming
   c) too high a firing rate
   d) wrong orifice
1654. How often should boiler tubes be cleaned?
   a) once a year
   b) once a month
   c) every two months
   d) as required
1655. How often should the low water cut-off be cleaned?
   a) annually
   b) monthly
   c) weekly
   d) every two years
1656. Boiler tubes are cleaned with:
   a) trisodium phosphate
   b) sodium chloride
   c) acid
   d) wire brush
1657. The highest operating pressure used on packaged Scotch boilers is:
   a) 30 lbs
   b) 50 lbs

c) 100 lbs
d) 150 lbs

1658. The diameter of the tubes on a high pressure boiler is:
    a) 1 to 2"
    b) 2 to 4"
    c) 4 to 6"
    d) none of the above

1659. How many times the working pressure is the pressure test for low pressure residential boilers?
    a) 1½ times
    b) 2 times
    c) 3 times
    d) none of the above

1660. The installation of a gas burner in a warm air furnace must include:
    a) low water cut-off
    b) hot air limit control
    c) "stack switch" combustion control
    d) circulating fan

1661. The purpose of a pressure regulator is to:
    a) increase the gas pressure
    b) maintain a constant gas pressure regardless of variations in gas supply
    c) prevent overheating the furnace
    d) provide a proper mixture of gas and air

1662. How is the oil pump coupling attached to the motor shaft?
    a) a woodruff key
    b) a flat key
    c) an Allen set screw
    d) a cap screw
    e) a machine screw

1663. How many controls does a gun-type oil burner have when used on a warm air system?
    a) one
    b) two
    c) three
    d) four
    e) five

1664. What voltages are found in most domestic gun-type oil burner systems?
    a) 24 and 120 volt
    b) 120 and 240 volt
    c) 24 and 230 volt
    d) 208 and 440 volt
    e) 24 and 220 volt

1665. Where is the primary control (bimetal type) mounted in a warm-air system?
   a) the plenum chamber
   b) the oil burner housing
   c) the furnace stack
   d) the cold air return
   e) at the thermostat
1666. Where is the relay located in a gun-type oil burner system?
   a) in the thermostat
   b) in the limit control
   c) in the primary control
   d) in the motor terminal box
   e) at the line switch
1667. The professional source from whom further information regarding the design and installation of ductwork may be found is:
   a) NESCA
   b) US Army Engineer Corps
   c) Encyclopedia of Heating Science
   d) Manual "B" of the Better Heating Bureau
1668. Closing the fingers on the diverter will adjust the:
   a) primary air
   b) safety pilot
   c) burner input
   d) pressure level
1669. One boiler horsepower is equal to supplying how many (EDR) square feet of steam radiation per hour?
   a) 140
   b) 200
   c) 180
   d) 233
1670. The load piping and pickup according to MCA standards is:
   a) 13%
   b) 25%
   c) 20%
   d) 30%
1671. How many square feet of heat transfer surface is equal to 1 boiler horsepower?
   a) 3
   b) 5
   c) 7
   d) 15
1672. If the input is 100,000 Btu, the efficiency is 80% and the 25% allowance is made for piping and pickup, what is the net capacity of the boiler?
   a) 80,000 Btu
   b) 40,000 Btu

c) 60,000 Btu
d) 20,000 Btu

1673. What type of burner is used with a Scotch boiler?
    a) vaporizing
    b) horizontal in-shot
    c) low-pressure
    d) rotary

1674. The area of the water surface at the water line is called:
    a) dis-engaging area
    b) heating surface
    c) steaming area
    d) blow-down area

1675. The ratio of heating surface to boiler horsepower is:
    a) 3 to 1
    b) 4 to 1
    c) 5 to 1
    d) 6 to 1

1676. All fire-tube boilers can produce what percent dry steam?
    a) 90%
    b) 94%
    c) 96%
    d) 98%

1677. What type of cleaning compound is used in starting up a boiler?
    a) trisodium phosphate
    b) sodium chlorate
    c) sulphur oxide
    d) lithium bromide

1678. Which of the following fuel oil(s) should generally be used with a high-pressure atomizing burner?
    a) No. 1 only
    b) No. 2 only
    c) No. 1 or 2
    d) none of the above

1679. Does UL require a motorized safety shut-off valve?
    a) yes
    b) no

1680. Which approval group required a pre-purge cycle?
    a) FIA
    b) FM
    c) UL
    d) all

1681. Does the pilot burn continuously?
    a) yes
    b) no

1682. What initiates the boiler control sequency?
    a) scanner
    b) timer
    c) limit control
    d) operating control
1683. Do industrial boilers always start on low fire?
    a) yes
    b) no
1684. The three principle types of boilers are:
    a) water-tube, HRT, locomotive
    b) cast iron, horizontal, vertical
    c) fire-tube, HRT, vertical
    d) water-tube, cast iron, fire-tube
1685. Scotch Marine boilers are used to develop pressures:
    a) below 250 psig
    b) above 250 psig
1686. Water-tube boilers operate with efficiencies up to:
    a) 70%
    b) 80%
    c) 90%
    d) 95%
1687. A boiler horsepower is equal to:
    a) 3.3475 Btu
    b) 33,475 Btu
    c) 334.75 Btu
    d) .33475 Btu
1688. Boiler horsepower is based on an evaporation temperature of:
    a) 230°F
    b) 220°F
    c) 224°F
    d) 212°F
1689. A boiler horsepower is equal to an evaporation rate of _____ pounds of steam per hour.
    a) 43.5
    b) 34.5
    c) 35.4
    d) 53.4
1690. What is the typical width of a ligament?
    a) ½"
    b) .75"
    c) 1"
    d) 1¼"
1691. What is the thickness of the tube sheet on low-pressure boilers?
    a) 3/16"

b) 5/16"
c) 7/16"
d) 9/16"

1692. Which boilers use the least refractory materials?
    a) 2 and 4 pass
    b) 3 pass
    c) 5 and 7 pass
    d) 6 pass

1693. How much insulation is applied to the shell?
    a) 1"
    b) 2 to 3"
    c) 3 to 4"
    d) 5"

1694. The ASME code for low-pressure boilers requires how many relief valves?
    a) 1 or more
    b) 2 or more
    c) 3 or more
    d) 4 or more

1695. What percent of low-pressure boilers are used in the commercial market?
    a) 50%
    b) 60%
    c) 75%
    d) 90%

1696. What certificate is necessary to obtain insurance?
    a) UL
    b) FM
    c) National Board Inspector
    d) FIA

1697. A column of water that exerts a pressure of one pound is:
    a) .433 ft high
    b) 3.95 ft high
    c) 2.31 ft high

1698. One Btu equals:
    a) 3412 ft-lb
    b) 2545 ft-lb
    c) 778 ft-lb

1699. An instrument used to determine specific gravity is a:
    a) hygrometer
    b) hydrometer
    c) viscosimeter

1700. The pour point of oil is the:
    a) congealing point
    b) point of slowest oil flow
    c) point where oil stops flowing

1701. An HRT boiler and a Wickes boiler are:
    a) both fired internally
    b) both fired externally
    c) not both fired the same way
1702. A siphon on a steam gauge is used to:
    a) reduce fluctuation of the gauge
    b) reduce excessive pressure
    c) eliminate false readings caused by high temperature
1703. A 300 psi spring on a safety valve can be set to operate at:
    a) 290 to 310 psi
    b) 285 to 315 psi
    c) 270 to 330 psi
1704. The oil burner requiring the highest oil pressure is the:
    a) steam atomizing
    b) air atomizing
    c) mechanical atomizing
    d) rotary cup
1705. Carbonate hardness is:
    a) permanent hardness
    b) temporary hardness
    c) chlorides
1706. If the belt driving a throttling flyball governor is slipping, the:
    a) engine speeds up
    b) cut-off will be later
    c) engine slows down
1707. A locomotive boiler:
    a) is not pitched
    b) pitches to the rear
    c) pitches to the front
1708. An HRT boiler expands:
    a) to the front
    b) to the rear
    c) downward
1709. Scale on the inside of water wall tubes causes:
    a) an interruption in water circulation
    b) high tube metal temperatures
    c) an increase in steam generation in the tubes
1710. Steam purification is accomplished by:
    a) line separators
    b) filters
    c) drum internals
1711. On a water strainer, the area of the holes in the strainer is:
    a) the same as the pipe area

b) greater than the pipe area
c) less than the pipe area

1712. Steam in a boiler is washed with in incoming feedwater to reduce the:
    a) moisture in the steam
    b) solids in the water
    c) solids in the steam

1713. Turbines used for boiler feed pumps are usually:
    a) single-stage impulse
    b) single-stage reaction
    c) multi-stage reaction

1714. The difference between AC and DC is:
    a) DC changes direction
    b) AC doesn't change direction
    c) AC changes direction and magnitude

1715. Valves between water column and boiler:
    a) are not permitted
    b) have non-rising stems
    c) are of OS&Y construction

1716. A pump pumps against:
    a) pressure
    b) a head
    c) velocity

1717. The discharge temperature of water from a blow-down tank should not exceed:
    a) 140°F
    b) 180°F
    c) 210°F

1718. On large HRT boilers, the heads are stayed with:
    a) through stays
    b) diagonal stays
    c) radial stays

1719. Engine knocks due to bearing wear can be quieted down by using:
    a) fuel oil
    b) light engine oil
    c) heavy engine oil

1720. On an automatic combustion control system, before soot blowing:
    a) have a high water level
    b) speed up the fuel supply
    c) place control system on manual or hand operation

1721. When loss of fire occurs on an automatic combustion control system, the dampers:
    a) open
    b) close
    c) stay the same

1722. The pump that is started with its discharge closed is the:
- a) rotary
- b) centrifugal
- c) reciprocating

1723. A firebox on a vertical firetube boiler is stayed with:
- a) radial stays
- b) an Adamson ring
- c) stay bolts

1724. The temperature of a Copes feedwater regulator is:
- a) higher at the top
- b) lower at the top
- c) constant throughout

1725. A non-return valve acts as a:
- a) globe valve
- b) angle valve
- c) check valve

1726. Hydrostatic lubricators are found on:
- a) steam inlet sides of engines
- b) steam inlet sides of turbines
- c) inlet sides of centrifugal pumps

1727. Wet atomizing steam to an oil burner:
- a) causes sparking
- b) increases amount of steam needed
- c) carbons up the burner tips

1728. Galvanize is made from:
- a) tin
- b) zinc
- c) copper

1729. If you had a boiler rated at 150 hp and an SWP of 200 psi, you would have a hard time maintaining steady flow at a pressure of:
- a) 50 psi
- b) 100 psi
- c) 180 psi
- d) 200 psi

1730. A conventional Sterling boiler has:
- a) three steam drums with a mud drum
- b) two steam drums with a mud drum
- c) two steam drums, one water drum and one mud drum

1731. The Detroit Roto-Stoker is:
- a) overfeed
- b) underfeed
- c) spreader
- d) pulverized fuel

1732. The Detroit Roto-Stoker needs:
    a) a windbox
    b) no windbox
    c) a large ash pit
1733. An 8 x 6 x 7 pump has a water piston with a diameter of:
    a) 8"
    b) 7"
    c) 6"
1734. The pressure a pump can work against is in proportion to the diameter of the:
    a) piston
    b) piston and length of the stroke
    c) piston, length of the stroke, and pump factor
1735. In the economic boiler, the short tubes are:
    a) smaller in diameter than the long tubes
    b) larger in diameter than the long tubes
    c) the same diameter as the long tubes
1736. In a closed heater:
    a) the pressure of the water is higher than the steam
    b) the pressure of the steam is higher than the water
    c) the water and steam pressures are the same
1737. The number of Btu in one gallon of #6 fuel oil is:
    a) 120,000 to 135,000
    b) 145,000 to 165,000
    c) 135,000 to 145,000
1738. An open heater is located on:
    a) the suction side of the boiler feed pump
    b) the discharge side of the boiler feed pump
    c) either side of the feed pump
1739. If an area of 1 square inch has pressure of 10 pounds, you would find the pressure on 1 square foot by multiplying by:
    a) 12
    b) 144
    c) 1728
1740. An increase in engine speed will cause boiler water level to:
    a) rise
    b) lower
    c) stay the same
1741. The steam valve on a simplex pump is thrown by:
    a) mechanical linkage
    b) steam
    c) pilot valve
1742. A single Bourdon tube:
    a) may be spiral wound

b) must be positioned so it can drain
c) may not be spiral wound
1743. A gear pump will have:
   a) no valves
   b) one suction and one discharge valve
   c) two suction and two discharge valves
1744. When two power boilers are connected to the same steam header, they must have a:
   a) (check) (stop) valve
   b) non-return valve
   c) non-return valve and a stop valve
1745. A globe valve to be used for boiler blow-down:
   a) may be used if of straight-away type
   b) may not be used
1746. Phenolphthalein is used to indicate:
   a) acidity
   b) sodium chloride
   c) alkalinity
1747. Chemically pure water has:
   a) high resistance
   b) low (reluctance) resistance
   c) none of the above
1748. Chemically pure water has:
   a) high conductivity
   b) low conductivity
   c) none of the above
1749. Low water level on an HRT boiler would first overheat:
   a) top tubes
   b) end of diagonal stays
   c) dry sheet
1750. A leaking safety valve will have:
   a) increased blowback
   b) decreased blowback
   c) none of the above
1751. Upon failure of a package boiler with electrically interlocked draft damper, the damper will:
   a) automatically close
   b) automatically open
   c) not be affected
1752. The efficiency of an injector being used as a pump is:
   a) 5%
   b) 10%
   c) 20%

1753. Fuses are connected in:
    a) series with the load
    b) parallel with the load
    c) series-parallel with the load
1754. A centrifugal pump has:
    a) one moving part
    b) two moving parts
    c) three moving parts
1755. When the counterpoise weight on a governor is increased, the engine speed:
    a) increases
    b) decreases
    c) remains the same
1756. What would be found in the line ahead of the fuel oil meter?
    a) a strainer
    b) a relief valve
    c) a pressure gauge
1757. If a safety valve is chattering, you would:
    a) increase blowback
    b) decrease blowback
    c) adjust spring tension
1758. The temperature limit of an alcohol thermometer is:
    a) 50°F
    b) 15°F
    c) 450°F
1759. The temperature limit of a mercurial thermometer is:
    a) 75°F
    b) 450°F
    c) 1100°F
1760. A manometer with one end open to atmospheric pressure reads:
    a) gauge pressure
    b) absolute pressure
    c) vacuum
1761. In a standard Sterling boiler, the steam outlet is on the:
    a) first drum
    b) second drum
    c) rear drum near feedwater inlet
1762. The minimum size of water gauge shut-off is:
    a) 1/2"
    b) 3/4"
    c) 5/8"
1763. The minimum size of water column connection to boiler is:
    a) 1/2"
    b) 3/4"
    c) 1"

1764. 23°F is equivalent to:
   a) -50° C
   b) 15° C
   c) 0° C
1765. On a standard Sterling boiler, there are water circulating tubes between:
   a) first and second drums
   b) second and rear drums
   c) all drums
1766. An inverted bucket trap can work with:
   a) steam pressure below rating
   b) steam pressure above rating
   c) any steam pressure
1767. An Ogee ring is used on:
   a) an Ogee boiler
   b) a Scotch boiler
   c) an upright boiler
1768. A device to measure high temperatures is called a:
   a) pyrometer
   b) manometer
   c) none of the above
1769. When joining threaded pipe, you put the compound on:
   a) the female thread
   b) the male thread
   c) both male and female threads
1770. Grease is composed of:
   a) vegetable oil and soap
   b) mineral oil and soap
   c) mineral oil and lye
1771. To clean a gummy commutator, use:
   a) carbona and gas
   b) oil and graphite
   c) none of the above
1772. The load on an engine is the:
   a) brake horsepower
   b) indicated horsepower
   c) actual horsepower
1773. The travel of the valve on an engine is:
   a) equal to the eccentricity
   b) twice the eccentricity
   c) half the eccentricity
1774. In a wet-type vertical boiler, the fusible plug is located:
   a) 2/3 of the way up on outside tube
   b) in upper tube sheet
   c) none of the above

1775. The closed feedwater heater has:
    a) a vent on water side
    b) a vent on steam side
    c) no vent
1776. An open feedwater heater:
    a) is vented to atmosphere
    b) has relief valve set at 15 lbs
    c) vacuum
1777. 1,000 watts is equal to:
    a) 34.5 Btu/min.
    b) 3413 Btu/hr
    c) 777.5 Btu/hr
1778. The Wickes and Manning boilers are:
    a) internally fired
    b) externally fired
    c) different in that one is internally fired and one is externally fired
1779. The piston and valve are going in the same direction at:
    a) cut-off
    b) release
    c) compression
    d) admission
1780. The inverted bucket trap has a:
    a) hole in the bucket to vent out air
    b) ball float
    c) none of the above
1781. More sulphur in fuel oil makes:
    a) more heat
    b) less heat
    c) more corrosion
1782. A simplex pump is usually used as:
    a) a vacuum pump or to pump air
    b) an oil pump
    c) a water pump
1783. The amount of coal burned per square foot of grate (hand fire, natural draft) is approximately:
    a) 20 lbs
    b) 35 lbs
    c) 50 lbs
1784. If packing gland is running hot:
    a) tighten packing evenly
    b) loosen packing evenly
    c) pour water on it
1785. The fan used to draw stack gases, etc. is:
    a) a forced draft fan

b) an induced draft fan
 c) none of the above
1786. To increase speed of engine, increase diameter of:
 a) driving pulley to governor
 b) driven pulley to governor
 c) idler pulley
1787. A pressure gauge may have:
 a) a Bourdon tube
 b) a diaphragm
 c) either
1788. Pressure gauges may be installed:
 a) in vertical position
 b) in horizontal position
 c) tilted forward for better visibility
1789. Grease fittings on modern machinery are:
 a) Alemite fittings
 b) Zerk fittings
 c) none of the above
1790. On a reciprocating pump, the diameter of the steam cylinder is:
 a) larger than the water
 b) smaller than the water
 c) the same as the water
1791. The bursting strength of steel is the:
 a) elasticity
 b) ultimate strength
 c) yield point
1792. In jacking an engine over by hand, it's good practice to have the steam valve:
 a) open
 b) closed
 c) cracked
1793. Which burner requires the lowest oil pressure?
 a) steam atomized
 b) air atomized
 c) rotary cup
1794. In an oil-burning system, the coarsest strainer would be on:
 a) suction side of pump
 b) discharge side of pump
 c) none of the above
1795. The amount of CO_2 in an efficiently operating oil burner boiler stack discharge is:
 a) 5% to 8%
 b) 10% to 14%
 c) 12% to 20%

1796. In a constant speed motor driving a pump, the pump speed may be raised by means of:
   a) hydraulic coupling
   b) Fordomatic
   c) none of the above
1797. A releasing Corliss engine governor is a:
   a) shaft governor
   b) loaded flyball governor
   c) pendulum governor
1798. The percentage of carbon in boiler steel is:
   a) 1.25%
   b) 0.25%
   c) 2.25%
1799. Silicon in boiler steel:
   a) is completely removed
   b) remains in trace amounts
   c) is undesirable
1800. The static head on a centrifugal pump varies:
   a) at the square of the speed
   b) directly at the speed
   c) inversely proportional to the speed
1801. A circular gauge glass will wear faster with a pH of:
   a) 11.5
   b) 8.5
   c) 6.5
1802. A circular gauge glass less than 18 inches long:
   a) needs no provision for expansion
   b) may have an automatic shut-off
   c) none of the above
1803. Making an adjustment for wear on the connecting rod at crankend:
   a) decrease clearance on head end
   b) decrease clearance on crank end
   c) shorten the connecting rod
1804. The type of coal that should be pulverized the finest is:
   a) anthracite
   b) sub-bituminous
   c) bituminous
1805. Torching in a recuperative air preheater is from:
   a) preheater plugging
   b) hole in plate between flue gas and air
   c) burning in preheater
1806. Inside admission piston valve with indirect valve linkage will be set for proper lead:
   a) 90 plus angle of advance behind the crank

b) 90 plus angle of advance ahead of crank
   c) 90 minus angle of advance behind crank
1807. Inside admission piston valve will be moving at release in the:
   a) same direction as piston
   b) opposite direction as piston
   c) none of the above
1808. A shaft governor with loose eccentric controls would have a:
   a) valve travel
   b) angle of advance
   c) throw of eccentric
1809. A generator with 4 poles is turning:
   a) 1200 rpm
   b) 1800 rpm
   c) 2400 rpm
1810. A generator with 6 poles turning at 1800 rpm produces current with:
   a) 60 cycles frequency
   b) 90 cycles frequency
   c) 120 cycles frequency
1811. Phosphorous in boiler steel:
   a) will make the steel hot short
   b) will make the steel cold short
   c) is a desirable quality
1812. Sulphur in boiler steel:
   a) will make the steel hot short
   b) is desirable
   c) will make the steel cold short
1813. The largest thermal drop will be through the:
   a) gas film
   b) tube metal
   c) scale
1814. The mud drum in a Heinie is at the bottom:
   a) of the front header
   b) inside the drum
   c) of the rear drum
1815. A safety valve is a:
   a) relief valve
   b) pressure safety valve to release pressure of the boiler
   c) none of the above
1816. On an accumulation test, the pressure should not rise more than:
   a) 3%
   b) 6%
   c) 10%
1817. A duplex pump, when steam bound, is full of:
   a) water

    b) water and steam
    c) steam
1818. A vacuum chamber on a duplex pump causes the pump to:
    a) race
    b) pulsate
    c) work smoothly at beginning and end of stroke
1819. A preheater is located:
    a) between boiler and economizer
    b) between stack and economizer
    c) on other side of stack
1820. The thickness of a manhole gasket is:
    a) 1/4"
    b) 1/2"
    c) 3/4"
1821. What has the highest volatile matter:
    a) coke
    b) bituminous coal
    c) anthracite coal
1822. On a fire-tube boiler, the:
    a) water is inside the tubes
    b) fire and gases pass through the tubes
    c) fire hits the outside of the tubes
1823. Fusible plugs are used on boilers of:
    a) 1200 psi
    b)   50 psi
    c) 1000 psi
1824. Two safety valves are used on boilers that have:
    a) more than 500 ft^2 of heating surface
    b) less than 500 ft^2 of heating surface
    c) less than 100 hp
1825. Flexible stay bolts are used on:
    a) locomotive type boilers only
    b) Heinie type boiler or water-tube
    c) firebox water legs boiler
1826. A dry pipe in a boiler is to:
    a) catch impurities
    b) catch the intrained condensate from the boiler
    c) make drier steam
1827. A pressure gauge corresponds to:
    a) the safety valve setting
    b) 5 times SWP
    c) 1½ to 2 times safety valve setting
1828. The lead on an engine is determined when:
    a) valve is in the mid position

b) eccentric is in the 90° position
c) crank is on dead center

1829. The type of turbine used on feedwater pump is:
   a) impulse
   b) reaction
   c) extraction

1830. The general procedure when warming up a turbine is:
   a) stop turbine
   b) turn over slowly
   c) turn over at maximum speed

1831. Caking coal is:
   a) low in oxygen
   b) high in oxygen
   c) average

1832. Lost motion in a duplex pump is:
   a) necessary
   b) unnecessary
   c) caused by wear

1833. To prevent water hammer:
   a) use superheated steam
   b) see right man that has hammer
   c) drain the lines

1834. A vacuum chamber is used on the:
   a) discharge side
   b) suction side
   c) chest cover

1835. An air preheater is used in conjunction with:
   a) an induced draft fan
   b) a forced draft fan
   c) an economizer

1836. A high-carbon coal is:
   a) bituminous
   b) anthracite
   c) lignite

1837. A blow-off cock should have plug held in place by:
   a) a gland and a nut
   b) no gland but with a nut
   c) a guard or gland marked in line on passage

1838. The speed of rotor on an impulse turbine should be:
   a) same speed or velocity of the steam
   b) 50% speed or velocity of the steam
   c) 87% speed or velocity of the steam

1839. Coal is of what origin?
   a) mineral

b) vegetable
c) animal

1840. A duplex pump in good working order has a suction lift of:
   a) 22 ft
   b) 26 ft
   c) 34 ft

1841. A pump in good working order should not have a slippage over:
   a) 5%
   b) 8%
   c) 10%

1842. A centrifugal governor is to:
   a) increase or decrease the steam pressure in the steam chest
   b) decrease the amount of steam flow
   c) increase the amount of steam according to the load on engine

1843. When the water in a boiler is dangerously low:
   a) speed up the pumps
   b) remove all combustible matter
   c) leave the feed pump and fuel supply alone

1844. When the water supply in a boiler is low:
   a) shut off the fuel supply
   b) restore the water level
   c) leave the pump alone

1845. Low level cut-off valves are used to shut off:
   a) fuel when water level is low
   b) water when water level is low
   c) air when water level is low

1846. In case of tube failure in a boiler:
   a) increase the forced draft
   b) open the doors
   c) increase induced draft

1847. To get complete combustion, excess air should be kept at:
   a) a maximum
   b) a minimum
   c) zero

1848. Power is measured by:
   a) Btu
   b) horsepower
   c) foot-pound

1849. Spontaneous combustion in stored coal is caused by coal high in:
   a) moisture and Sulphur
   b) hydrocarbons
   c) fixed carbons

1850. The water and steam in a Copes feedwater regulator:
   a) move rapidly

b) move slowly
c) remain motionless

1851. The water at the bottom of the water glass of an HRT boiler should be:
    a) 3" above the top row of tubes
    b) 4" above the top row of tubes
    c) 6" above the top row of tubes

1852. The safety valve should be set when the:
    a) water level is high
    b) water level is low
    c) drum is drained

1853. Lead is found on an engine when:
    a) the piston is on dead center
    b) the valve is in mid-travel
    c) the piston is in mid-travel

1854. The frequency most frequently used is:
    a) 0 cycles
    b) 50 cycles
    c) 60 cycles

1855. Soda ash is used to:
    a) remove scale from drum
    b) alkalize water
    c) remove temporary hardness

1856. When temperature and pressure increase, the latent heat:
    a) increases
    b) decreases
    c) stays the same

1857. If a pump is vapor-bound, it:
    a) stops
    b) short strokes and is noisy
    c) races

1858. A Wickes vertical boiler is supported by:
    a) lugs on lower drum
    b) steam drum
    c) mud ring

1859. An oil separator is used on the exhaust line of:
    a) an engine
    b) an oil heater
    c) a turbine

1860. Caking coals are burned on:
    a) an underfeed stoker
    b) a traveling grate
    c) a Detroit

1861. Correct preheated air in stoker firing:
    a) slows firing rate

b) aids combustion
  c) prevents slag formation
1862. Vertical water-tube boilers are supported by:
  a) lugs on lower drum and allowed to expand upwards
  b) lugs on lower drum and slings connected to top drum
  c) hangers
1863. A dash pot on a governor:
  a) helps close valve on Corliss engine
  b) prevents hunting
  c) helps change speed
1864. An evaporator is used to:
  a) distill make-up water
  b) take impurities out of oil
  c) leave air out of open heater
1865. Boiler feedwater can be:
  a) above 212°F
  b) below 212°F
  c) at 212°F
1866. Which is not a positive pump?
  a) reciprocating
  b) rotary
  c) centrifugal
1867. The drum of a straight tube box-header boiler is:
  a) parallel with tubes
  b) parallel with header
  c) perpendicular to tubes
1868. Boiler hp, according to most city codes, is:
  a) 34.5 lbs/hr at 212°F
  b) 10 ft^2 heating surface
  c) 33,000 Btu
1869. Water with an excess of hydroxyl ions in solution will be:
  a) acid
  b) neutral
  c) alkaline
1870. Continuous blow-down:
  a) removes oil
  b) removes impurities on surface water
  c) controls accumulation
1871. Pump pressure in a globe valve:
  a) is above seat
  b) is below seat
  c) has no pressure
1872. Oil burners serve to:
  a) vaporize oil

b) atomize oil
c) burn the oil
1873. Removing irregular-size coal causes:
    a) more uniform combustion
    b) better firing
    c) an even fuel bed
1874. Which of these would be more difficult to keep from smoking?
    a) coking coal
    b) anthracite
    c) bituminous
1875. Gun-type burners (such as Troy, Peabody, etc.) should be:
    a) left in burner
    b) removed, cleaned and hung on torch rack
    c) cleaned and left in burner
1876. Transformers are used to:
    a) increase or decrease voltage
    b) lessen transmission losses
    c) change AC to DC
1877. A commutator is found on:
    a) an alternating current motor
    b) a DC motor
    c) a transformer
1878. Which of these has the highest O_2 content?
    a) anthracite coal
    b) bituminous coal
    c) semi-anthracite coal
1879. The flash point, in comparison to ignition, is:
    a) above
    b) even
    c) less
1880. In starting a large motor, you use:
    a) rheostat
    b) transformer
    c) exciter
1881. In cleaning electrical parts, use:
    a) gasoline
    b) emery cloth
    c) sandpaper
1882. Which of these motors uses an exciter?
    a) synchronous
    b) induction
    c) squirrel cage
1883. A synchronous motor uses:
    a) AC

b) DC
   c) AC and DC
1884. Pump pressure on a globe valve is:
   a) above seat
   b) below seat
   c) no pressure on seat
1885. Frequency most often used in AC is:
   a) 0 cps
   b) 60 cps
   c) 50 cps
1886. Piston rod is connected to connecting rod by the:
   a) crosshead
   b) guide
   c) suitable pin
1887. A column of water 1 square inch x 1 foot high equals .433 psi. A column of water 5 square inches x 1 foot high has pressure at bottom of:
   a) .433 psi
   b) 2.31 psi
   c) 105 psi
1888. A duplex pump and centrifugal pump, under same conditions, are:
   a) just as effective
   b) the duplex is more effective
   c) the centrifugal is more effective
1889. Water space in a vertical firetube boiler is less:
   a) for more steam space
   b) to superheat system
   c) for less carryover
1890. The entrance to a blow-down tank points:
   a) downward
   b) upward
   c) straight in
1891. All fire-tube HRTs are:
   a) externally fired
   b) internally and externally fired
   c) internally fired
1892. Locomotive boilers are fired:
   a) externally
   b) internally
   c) both
1893. Blow-down tanks are vented to:
   a) atmosphere
   b) sewer
   c) not vented

1894. Finned tubes on water walls are to:
   a) protect refractory
   b) extract more radiant heat
   c) put maintenance cost on furnace settings
1895. On the water end, there are at least:
   a) 8 valves on a duplex pump
   b) 10 valves on a duplex pump
   c) 12 valves on a duplex pump
1896. A Copes is a:
   a) thermostatic mechanical control
   b) thermostatic pressure control
   c) thermostatic vapor control
1897. A duplex pump has:
   a) no lap
   b) lap
   c) little lap
1898. The heat added to water at 32°F to bring temperature to 212°F is:
   a) sensible heat
   b) latent heat
   c) specific heat
1899. The heat added to water at 212°F to make steam at 212°F is:
   a) sensible heat
   b) latent heat
   c) specific heat
1900. The minimum size of pipe on column to water glass is:
   a) 1"
   b) 3/4"
   c) 1 1/2"
1901. The minimum size of drain on water column is:
   a) 1"
   b) 3/4"
   c) 1/2"
1902. Most large factories use:
   a) one-phase motors
   b) two-phase motors
   c) three-phase motors
1903. On a hot water heating system, there is a:
   a) relief valve
   b) spring-loaded pressure safety valve
   c) spring-loaded relief valve
1904. If an electric motor failed to start, you would first look at the:
   a) fuses
   b) rheostat
   c) switch

1905. A pH of 7 in boiler water indicates:
   a) alkalinity
   b) acidity
   c) neutral
1906. Fusible plugs on standard Sterling boilers are found in:
   a) front drum
   b) middle drum
   c) mud drum
1907. Shaft governors employ:
   a) centrifugal force
   b) centrifugal and inertia force
   c) inertia force
1908. The temperature at which a "D" slide valve is limited to is dependent upon:
   a) the warping of the valve
   b) lubrication of the valve
   c) action of the valve
1909. The piping connections allowed between boiler and water column are:
   a) damper regulator and steam gauge only
   b) none whatever
   c) anything that does not allow an appreciable amount of steam flow
1910. Lead is given an engine to:
   a) compensate for angularity
   b) help the piston get started
   c) cause compression
1911. Water softener:
   a) removes impurities
   b) makes water caustic
   c) makes water acid
1912. The lowest part of the water glass should be at least how many inches above the lowest safe level?
   a) 1"
   b) 2"
   c) 3"
1913. If the hi-low water alarm is whistling:
   a) blow-down boiler
   b) determine the true water level
   c) bank boiler
1914. Valves between column and boiler should be:
   a) non-rising stem gate valve
   b) outside screw and yoke type
   c) globe
1915. Which is the hardest scale?
   a) calcium

b) magnesium
c) silica

1916. Bearings on an engine are composed of:
   a) brass
   b) bronze
   c) babbit

1917. A third class stationary engine license covers:
   a) 5000 ft² heat and 20 hp
   b) 7500 ft² heat surface and 50 hp
   c) 7500 ft² heat surface and 100 hp

1918. A pH of 6 is:
   a) alkaline
   b) acid
   c) neutral

1919. A pH of 8 is:
   a) alkaline
   b) acid
   c) neutral

1920. Fusible plugs melt at:
   a) 450°F
   b) 350°F
   c) 600°F

1921. Temperature in a furnace is:
   a) 1600°F
   b) 2500°F
   c) 3600°F

1922. The length of a valve affects:
   a) lead
   b) cutoff
   c) angle of advance

1923. A compound engine always has:
   a) two cylinders
   b) two cylinders or more
   c) two stages

1924. When starting an engine, the piston is at:
   a) head end dead center
   b) crank end dead center
   c) mid-position

1925. The energy used in generating electricity is:
   a) mechanical
   b) chemical
   c) friction

1926. Tighten packing on piston rod when the engine is:
   a) stopped

b) running
c) leaking excessively

1927. When warming up an oil-fired boiler, you start the:
    a) top burner
    b) side burner
    c) lower burner

1928. When starting fuel oil, fill the tank to:
    a) near capacity
    b) near capacity in summer, capacity in winter
    c) capacity

1929. When trammeling an engine, you:
    a) find mid-position
    b) find head and crank end dead center
    c) square the valve

1930. Oil in air compressor should be:
    a) high flash point
    b) low flash point
    c) SAE 60

1931. On an impulse-driven turbine, the fixed blades and steam jets are connected to the:
    a) shaft
    b) rotor
    c) casing

1932. Which of these three is a good electrical conductor?
    a) water
    b) leather
    c) rubber

1933. One cubic foot of water from the river weighs:
    a) 62.4 lbs
    b) 64.8 lbs
    c) 57 lbs

1934. When a motor runs faster, friction hp is:
    a) the same
    b) more
    c) less

1935. On DC, watts are equal to:
    a) volts x amps
    b) volts x ohms
    c) volts + amps

1936. The efficiency of a turbine is:
    a) 20%
    b) 25%
    c) 30%

1937. As pressure rises, the boiling point goes:
   a) down
   b) up
   c) stays the same
1938. Tubes are beaded in fire-tube boilers:
   a) to support tube sheet
   b) so tube ends will not burn off
   c) for better gas circulation
1939. Tubes in a water-tube boiler are:
   a) mostly beaded
   b) mostly flared
   c) flared and beaded
1940. On fire-tube boilers, tubes are:
   a) flared
   b) flared and beaded
   c) beaded
1941. The best way to prevent spontaneous combustion is:
   a) increase O_2
   b) decrease O_2
   c) leave the same
1942. Greatest loss in a boiler is:
   a) dry gases up stack
   b) leaks in setting
   c) unburned coal in ashes
1943. An injector has:
   a) high thermal efficiency
   b) economical way to feed water to boiler
   c) higher suction left than pump
1944. A 350 hp boiler used for heating is down for 4 weeks. The best action is:
   a) leave banked fire with slight pressure in it
   b) take out of service, using all service precautions
   c) just take out of service
1945. What will cause flash-back when starting an oil burner?
   a) mixture of oil vapor, air and ignition
   b) too much oil
   c) too much air
1946. A standard Sterling boiler has:
   a) 2 flue passages
   b) 3 flue passages
   c) 4 flue passages
1947. After renewing water glass, open:
   a) steam valve first
   b) water valve first
   c) both at the same time

1948. Coke breeze is burned by:
- a) an underfeed stoker
- b) a chain grate (grate bar type)
- c) a spreader stoker

1949. Electric fuse plugs are marked in:
- a) volts
- b) amperes
- c) watts

1950. Height of water in an air chamber of a reciprocating pump is:
- a) 3/4 full
- b) 1/2 full
- c) 1/4 full

1951. What causes greatest heat losses?
- a) blowing down boiler
- b) carbon dioxide
- c) carbon monoxide

1952. A vertical fire-tube boiler has:
- a) submerged flues and dutch oven
- b) enclosed firebox and submerged flues
- c) exposed tubes and dutch oven

1953. When changing from forced to induced draft, use:
- a) a larger fan
- b) a smaller fan
- c) same fan

1954. A lever on a safety valve is used to:
- a) raise valve off seat
- b) release boiler pressure
- c) both at the same time

1955. Simplex and duplex pumps are:
- a) horizontal only
- b) vertical only (grate bar type)
- c) horizontal or vertical depending on service used for

1956. Fire cracks on an HRT are at:
- a) girth seam
- b) front head seam
- c) rear head seam

1957. How much larger fan is required for induced fan than a forced fan?
- a) 50%
- b) 25%
- c) same size

1958. Chain grates are:
- a) grate units attached on bars
- b) continuous chain conveyor
- c) grate bars forming chain

1959. What will cause a crack in a furnace wall?
   a) starting a fire too fast
   b) starting with a flash
   c) poor combustion and draft
1960. Steam vapor is:
   a) visible
   b) invisible
   c) liquid
1961. Which pump, under same conditions, is most effective?
   a) reciprocating duplex
   b) rotary
   c) centrifugal
1962. Which pump uses most steam?
   a) reciprocating
   b) simplex
   c) compound
1963. In balanced draft:
   a) a forced draft fan puts in more air than induced fan
   b) both fans are equal
   c) an induced fan puts in more air than a forced fan
1964. Stoker of the grate bar key type is:
   a) underfeed
   b) chain grate
   c) traveling grate
1965. A slide valve engine is considered to be:
   a) high speed
   b) low speed
   c) medium speed
1966. A steam engine with two steam cylinders has to be:
   a) compound
   b) not necessarily compound
   c) conventional
1967. Safety valves pop to release saturated boiler steam at what range:
   a) 3% above safe working pressure
   b) 6% above safe working pressure
   c) 10% above safe working pressure
1968. Litmus paper is used to test for:
   a) acid
   b) alkaline
   c) both A and B
1969. The type of pump used for pumping heavy oil is a:
   a) rotary
   b) centrifugal
   c) reciprocating

1970. Which has a coking arch?
   a) chain grate
   b) traveling grate
   c) underfeed
1971. Which of these are hardest scaled?
   a) calcium
   b) carbonates
   c) silicates
1972. A flame rod in oil firing:
   a) lights off a burner
   b) is a safety device if flame goes out
   c) is for use as a torch
1973. What affects natural water circulation in a boiler?
   a) outgoing steam velocity
   b) feedwater coming in
   c) difference of density of steam and water
1974. For steam passing through stationary blades of an impulse turbine, the velocity:
   a) increases
   b) decreases
   c) stays the same
1975. A duplex pump valve has:
   a) lap
   b) no lap
   c) uses steam expensively
1976. Pressure used on a "D" slide valve is:
   a) 100 to 200 lbs
   b) 200 to 300 lbs
   c) 300 to 400 lbs
1977. Fuel oil has:
   a) more Btu than anthracite coal
   b) less Btu than anthracite coal
   c) the same Btu as anthracite coal
1978. Flange gaskets on high pressure steam are:
   a) thin
   b) medium
   c) thick
1979. Which valve restricts flow the least?
   a) globe
   b) gate
   c) petcock
1980. Dry pipe in a boiler:
   a) is open at both ends

b) has small hole on top
c) has small hole in bottom
1981. In balanced draft, pressure in furnace is compared to atmospheres:
    a) higher
    b) lower
    c) the same
1982. In a traveling grate, you burn:
    a) low-volatile coal
    b) high-volatile coal
    c) high-volatile coking coal
1983. The minimum blowback on a safety valve is:
    a) 2 to 8 lbs
    b) 6 lbs
    c) 10 lbs
    d) never over 4% of pressure
1984. Soot blowers blow:
    a) between tubes
    b) on tubes
    c) toward stack
1985. An Adamson ring is used on:
    a) Scotch Marine boilers
    b) water-tube boilers
    c) a Manning boiler
1986. An ogee ring is used on:
    a) Scotch Marine boilers
    b) water-tube boilers
    c) Manning boilers
1987. Which wire offers best resistance to flow?
    a) conductor
    b) non-conductor
    c) standard wire
1988. A safety valve on a long pipe would:
    a) note seat
    b) open easier
    c) open later
    d) chatter
1989. When an engine gets a sudden load, it:
    a) speeds up
    b) slows down
    c) runs the same
1990. You get more superheated steam by:
    a) adding more moisture to steam
    b) adding more pressure to steam
    c) throttling

1991. Current load is measured by:
- a) an amp meter
- b) a watt meter
- c) a volt meter

1992. In an oil-fired boiler, when boiler is lighting off, there is:
- a) no smoke
- b) a light brown haze
- c) gray smoke

1993. If a chimney is smoking heavily, there is:
- a) a lack of O_2
- b) too heavy a fuel bed
- c) the wrong fuel

1994. An air chamber on a duplex pump can be found on:
- a) the suction side
- b) the discharge side
- c) both sides

1995. A horizontal "D" slide valve engine is set for equal:
- a) lead
- b) cut-off
- c) lead and cut-off

1996. If a safety valve was leaking, you would:
- a) tighten down and spring until leaking stops
- b) close the stop valve
- c) remove from boiler and repair

1997. When lighting off a boiler with an internal superheater, you would:
- a) open drain
- b) by-pass the superheater
- c) shut off the superheater

1998. How much air is needed for combustion?
- a) 12 lbs
- b) 15 lbs
- c) 18 lbs

1999. A minimum blow-off line is:
- a) 1"
- b) 2 1/2"
- c) 3/4"

2000. Which boiler has a blow-down tank?
- a) a portable boiler
- b) a low pressure boiler
- c) a power boiler

2001. A return trap feeds:
- a) condensate only
- b) make up water and condensate
- c) power boiler

2002. On a hydrostatic lubricator, the sight glass is full of:
   a) water
   b) half water and half oil
   c) oil
2003. On an oil or gas-fired boiler, in case of serious over pressure:
   a) lift safety valve by hand
   b) shut off fuel and lift safety valve by hand
   c) shut off fuel
2004. Water can evaporate at:
   a) 212° only
   b) above 212° only
   c) any temperature
2005. Which valve has the least resistance?
   a) gate
   b) globe
   c) right angle check
2006. Which valve is easiest to tell when open?
   a) gate
   b) globe
   c) OS&Y
2007. When the centrifugal pump is running, what stops water from going back to the suction side?
   a) casing
   b) impeller
   c) wearing ring
2008. You purge a boiler of gas when:
   a) starting
   b) shutting down
   c) operating
2009. If superheated, the volume of a given weight of steam:
   a) increases
   b) decreases
   c) stays the same (C.E. 19.2)
2010. Which contains the most Btu per volume?
   a) gas
   b) oil
   c) coal
2011. Acid-forming substances are removed from feedwater by:
   a) chemicals
   b) heat
   c) filtration
2012. With traveling grate stokers or chain grates, ignition should become started and stable after entering furnace:
   a) as soon as possible after entering boiler

    b) less than ¼ distance after entering
    c) about ¼ distance after entering
2013. A manometer reads 2.8" of water. This corresponds to what pressure?
    a) 14.6
    b) 14.8
    c) 14.7
    d) 11.9
2014. Lead affects the following events in what order?
    a) admission, cut-off, release and compression
    b) admission, release, cut-off and compression
    c) admission, compression, exhaust and cut-off
2015. Fire-tube boilers are usually not used above:
    a) 150 lbs
    b) 300 lbs
    c) 450 lbs
2016. In case of low water in a Scotch Marine boiler, what is damaged first?
    a) crown sheet
    b) top-row tubes
    c) flues
2017. If a boiler is to be washed out, it is:
    a) washed while boiler is hot
    b) allowed to set and dry out
    c) washed as soon as possible after emptying
2018. Entrance to the blow-down tank should be:
    a) not less than blow-down pipe from boiler
    b) 1½ times larger than blow-down pipe from boiler
    c) 2 times larger than blow-down pipe from boiler
2019. When using an extension light inside a boiler, it:
    a) should be grounded
    b) should not be grounded
    c) doesn't make any difference
2020. Which is proper procedure in blowing-down a boiler?
    a) intermittently
    b) continuously
    c) spasmodically
2021. A compound gauge:
    a) measures pressure and vacuum
    b) has two Bourdon tubes
    c) shows absolute and gauge pressure
2022. In a centrifugal pump, which way does the impeller curve?
    a) backwards to the rotation of the shaft
    b) backwards against the shaft
    c) they are straight

2023. Fusible plugs are changed:
   a) every year
   b) every 5 years
   c) every 10 years
2024. When water leaves a centrifugal pump, it will have a:
   a) high velocity and low pressure
   b) high pressure and low velocity
   c) high pressure and high velocity
2025. Superheater fins are made of:
   a) steel
   b) cast iron
   c) brass
   d) aluminum
2026. An elliptical manhole plate runs:
   a) longitudinally
   b) radially
   c) with girth seam
2027. In burning breeze fuel, it would burn best on:
   a) chain grate
   b) spreader
   c) underfeed
2028. Where is most heat lost on a boiler?
   a) too much O_2
   b) too much H
   c) up the stack
2029. A simplex pump is:
   a) single-acting
   b) double-acting
   c) both
2030. To tighten up a screwed pipe, you would use a:
   a) crescent wrench
   b) stillson wrench
   c) monkey wrench
2031. Pipe threads:
   a) have a taper of 70
   b) are always tapered to keep threads tight
   c) are not tapered (as with a machine bolt)
2032. The greatest variety of coal can be burned on:
   a) an underfeed
   b) a spreader
   c) a chain grate
2033. Fittings on a steam pipe are made of:
   a) steel

b) cast iron
c) carborundum
2034. Flange joints are best suited because they:
   a) have no gaskets
   b) are easy to maintain
   c) have flexibility
2035. What would cause soft sludge to become scale?
   a) temperature is raised
   b) feedwater treatment is discontinued
   c) improper feedwater treatment
2036. As a boiler deteriorates, the safety factor:
   a) goes up
   b) goes down
   c) stays the same
2037. With a magnetic valve, there is circulation:
   a) when valve closes
   b) when valve opens
   c) at all times
2038. With an induction motor, which is stationary?
   a) rotor
   b) stator
   c) commutator
2039. At what part of stroke does steam cut off on a pump?
   a) 1/3
   b) 1/2
   c) 3/4
2040. On a steam engine that has no lap or lead, cut-off occurs:
   a) at or near end of stroke
   b) at 3/4 of stroke
   c) at 5/8 of stroke
2041. On a common slide valve engine, cut-off occurs:
   a) 1/4
   b) 5/8
   c) 7/8
2042. On a streamline, strainer is:
   a) before trap
   b) behind trap
   c) in trap
2043. On a fuel-oil system, pump should be protected by a strainer:
   a) before pump
   b) after pump
   c) before and after pump
2044. A non-return trap on line that is to drain is placed:
   a) above line

b) below line
c) equal with line
2045. Excess oil in an oil-fired system returns to the:
a) tank
b) suction side of pump
c) sewer
2046. A hydrostatic lubricator on an engine is to lubricate the:
a) crank
b) crosshead
c) steam chest
2047. A small vertical steam engine crank is lubricated by:
a) a splash system
b) a forced oil pump
c) gravity
2048. On a gas-fired system, the magnetic valve is held open by:
a) gas pressure in line
b) electricity
c) gas pressure keeps it open, electricity closes it
2049. If a boiler is foaming badly, what would you do first?
a) blow-down boiler
b) determine true water level
c) open safety valve by hand
2050. On a throttling governor, steam cut off from engine:
a) reduces steam pressure in steam chest
b) alters cut-off
c) shuts off steam from engine
2051. A non-return valve acts as a:
a) check valve
b) relief valve
c) safety valve
2052. The steam gauge on a boiler operating at 100 psi should register pressures up to:
a) 100 lbs
b) 150 lbs
c) 200 lbs
2053. What is the vacuum in an injector?
a) 14"
b) 11 ½'
c) 9"
2054. A plunger pump is packed:
a) inside
b) outside
c) both inside and outside

2055. What type of stoker handles any size coal best?
   a) spreader
   b) underfeed
   c) chain grate
2056. How does an expansion trap work?
   a) heat of steam
   b) when water cools off
   c) mechanical
2057. On a simplex pump, the steam valve is:
   a) mechanically operated
   b) operated by pilot valve
   c) steam thrown
2058. A safety device on a line to prevent overload is:
   a) an open switch
   b) a closed switch
   c) a circuit breaker
2059. On a duplex pump, the steam valve is:
   a) steam operated
   b) mechanically operated
   c) operated by pilot
2060. In feedwater treatment, ppm means:
   a) pounds per minute
   b) parts per minute
   c) parts per million
2061. Water temperature to suction of boiler feed pump should be:
   a) 180°F
   b) 200°F
   c) not hot enough to cause the pump to become vapor-bound
2062. An acceptance test on a new turbine should be made after it enters commercial service within:
   a) 2 weeks
   b) 2 months
   c) 1 year
2063. When testing a pressure gauge:
   a) use a gauge of similar type
   b) raise pressure until safety is about to pop
   c) use a dead weight testing applicator
2064. In cleaning a crank case or lubricator oil tank, you must use:
   a) waste material
   b) wiping cloth
   c) steel wool
2065. On a non-return trap, what opens the discharge valve?
   a) heat of steam

b) cooling effect of water
c) it is mechanically operated

2066. An automatic injector will restart itself when there is an interruption in:
a) water
b) steam
c) steam or water

2067. Filtration is a:
a) sedimentation process
b) mechanical process
c) chemical process

2068. Filter material is most often found in:
a) an open heater
b) a discharge line from pump
c) a suction line from pump

2069. Closer blow-down attention (in similar conditions) must be given to:
a) internal treatment
b) external treatment
c) both a and b, depending on the system used

2070. On a Copes feedwater regulator, the:
a) water level cannot be changed readily
b) water level can be changed readily
c) regulator has a by-pass

2071. A Copes:
a) should be installed at a 45° angle
b) is operated by difference in temperature of water and steam
c) is operated by a difference in pressure

2072. The factor in the combustion on a stoker grate that reduces or limits the combustion rate is the temperature:
a) of flue gases
b) of preheated air
c) at which fuel ignites

2073. When a boiler is coal, the type of soot blower to use is:
a) steam hand lance
b) regular steam blower
c) compressed air blower

2074. Valves or horizontal pipe, when possible:
a) stem points upward
b) stem points downward
c) elbow

2075. Highly contaminated steam use:
a) open heater
b) closed heater
c) dump to sewer

2076. A safety valve seat sticks less often with a seat of:
   a) 45°
   b) 90°
   c) flat
2077. When you have wet steam on an oil burner atomizer, it:
   a) causes intermittent firing
   b) carbons the tip
   c) aids combustion
2078. Which contains the most water?
   a) saturated steam
   b) superheated steam
   c) wet steam
2079. On a locomotive boiler. the hot gases pass:
   a) under the drum and through tubes to stack
   b) through furnace then through tubes to stack
   c) through furnace then under tubes to stack
2080. Which oil burner do you start off first?
   a) center
   b) top
   c) bottom
   d) sides
2081. To air-clean electrical equipment, use:
   a) low moisture
   b) high moisture
   c) oxygen
2082. Very large HRT boilers have:
   a) 1 manhole
   b) 2 manholes
   c) 3 manholes
2083. A high velocity of air in oil firing causes:
   a) the flame to spark
   b) high combustion
   c) high rating
2084. An HRT is a:
   a) one-pass boiler
   b) two-pass boiler
   c) three-pass boiler
2085. If you decrease the lost motion on a duplex pumo, the stroke is:
   a) increased
   b) decreased
   c) stays the same
2086. If a valve was stamped SWP 125, that would be for:
   a) any fluid or gas

b) water but not steam
c) steam but not water
2087. Underfeed stokers carry a fuel bed of:
   a) 6" to 12"
   b) 2" to 6"
   c) 2" or less
2088. Which can burn the most coal per square foot in grate surface per hour?
   a) chain grate
   b) single retort stoker
   c) hand firing
2089. When you have a temporary drop in load with a mechanical oil burner:
   a) adjust fuel supply
   b) change burner tips
   c) adjust air
2090. Where is the greatest loss in an electrical wire?
   a) in amperes
   b) in ohms
   c) in volts
2091. Which engine has more parts which can wear?
   a) a tandem engine
   b) a single-cylinder engine
   c) a cross-compound engine
2092. What prevents steam pressure in a steam turbine chest from getting too high?
   a) a relief valve
   b) a governor
   c) a throttle
2093. What type of trap would you use on high pressure for greatest effect?
   a) thermostatic
   b) ball float
   c) inverted bucket
2094. What would cause a motor to vibrate?
   a) motor too small
   b) bearings out of line
   c) rotor out of balance
2095. When you start a rotary pump, you are supposed to leave discharge:
   a) open
   b) closed
   c) partly open
2096. A non-return valve can be:
   a) double-acting
   b) single-acting
   c) both
2097. Which water will cause most scale?
   a) inside water treatment

b) zeolite
c) both the same
2098. When bottom valve on water glass is plugged up, the water:
   a) rises slowly
   b) rises fast
   c) stays the same
2099. When upper valve on water glass is plugged up, the water:
   a) rises slowly
   b) rises fast
   c) stays the same
2100. The percent of overspeed trip-off turbine is:
   a) 1%
   b) 5%
   c) 10%
2101. The blow-down tank outlet is:
   a) the same size as blow-down line
   b) 1½ times as large as blow-down line
   c) 2 times as large as blow-down line
2102. On an engine with a flywheel, the flywheel is used to:
   a) balance the engine
   b) drive a pulley
   c) absorb energy from the engine
2103. When you have highly contaminated exhaust steam, use:
   a) open heater
   b) closed heater
   c) barometric
2104. What type of fuse has a changeable link?
   a) plug
   b) cartridge
   c) fusetron
2105. When an engine develops more power, friction horsepower:
   a) increases
   b) decreases
   c) stays the same
2106. Water corrodes most with what kind of pipe?
   a) copper
   b) steel
   c) cast iron
2107. You would vent boiler when draining:
   a) at atmospheric pressure
   b) below atmospheric pressure
   c) before draining
2108. The container used to store mercury should be made of:
   a) chamois

b) copper
c) glass or earthenware
2109. To tighten the hex nut on the shoulder of a valve that is leaking, use:
   a) a stillson wrench
   b) an Allen wrench
   c) a monkey wrench
2110. The fuel bed in a traveling grate is:
   a) over 6"
   b) 2" to 6"
   c) 2" even
2111. Water in an open heater is:
   a) at the boiling point
   b) above the boiling point
   c) lower than the boiling point
2112. When blowing down a boiler and not in sight of water glass, you would:
   a) station another man to watch column
   b) drain boiler
   c) blow-down until low water alarm blows
2113. The easiest method of putting pipe together is:
   a) screwed on
   b) bolted on
   c) welded together
2114. The first thing an engineer does when taking over a shift is to check the:
   a) log book
   b) safety valve
   c) water column
2115. If the top water glass was plugged, the water would:
   a) stay the same
   b) rise quickly to the top
   c) rise slowly to the top
2116. How do you know if an engine is out of line?
   a) excessive use of steam
   b) it will not handle overload
   c) connecting rods or guides are wobbly
2117. What types of stays are in Scotch Marine boilers?
   a) through head to head
   b) diagonal stays
   c) hollow stays
2118. The heat of combustion gases passing through boiler:
   a) radiation
   b) conduction
   c) convection
2119. A superheater, when working, is filled with:
   a) saturated steam

b) dry steam
c) water
2120. Using feedwater to wash outgoing steam helps to:
   a) reduce moisture in steam
   b) reduce the amount of solid concentration
   c) heat feedwater
2121. Butting or attaching another safety valve that is leaking is:
   a) dangerous
   b) permissible
   c) all right if pressure doesn't rise above maximum
2122. Increasing the area of the underside of the valve disc on a safety valve will cause it to:
   a) increase blow-back
   b) suddenly pop
   c) slowly open
2123. To remove lime coating on inside of injector, use:
   a) muriatic acid
   b) sulphuric acid
   c) a solution of muriatiac acid
2124. Which of these is vented to atmosphere?
   a) open heater
   b) closed heater
   c) both
2125. Flash point of oil compared to ignition is:
   a) 80° less
   b) 50° less
   c) 20° less
2126. If water glass should break, you would:
   a) shut off steam valve first
   b) shut off water valve first
   c) drain boiler
2127. As feedwater, which would contain the most impurities?
   a) water being internally treated
   b) water being externally treated
   c) raw water
2128. A Scotch boiler is:
   a) a fire-tube
   b) a water-tube
   c) an inclined tube
2129. In electrical, resistance is:
   a) resistor
   b) ohms
   c) volts

2130. Holes in a strainer, in relation to pipe size, should be:
    a) less
    b) slightly more
    c) equal to
2131. Excess steam pressure is removed from turbine casing by:
    a) a relief valve
    b) a governor
    c) a condenser
2132. A Scotch Marine boiler is of what type?
    a) horizontal
    b) vertical
    c) locomotive
2133. In drying a boiler after cleaning, use:
    a) quick lime
    b) soda ash
    c) solution of lime
2134. The flash point of a lubrication oil used in a steam cylinder is:
    a) when it holds a steady flame
    b) above fire point
    c) below fire point
2135. Tube ends of water walls usually terminate in:
    a) mud drums
    b) cross boxes
    c) headers
2136. Which of these is not considered carryover?
    a) slugs of water
    b) 80% steam
    c) saturated steam
2137. The connection between piston rod and connecting rod is a:
    a) crank pin
    b) cross head
    c) connecting rod
2138. A crane uses a:
    a) vertical boiler
    b) horizontal boiler
    c) water-tube boiler
2139. Which motor is the largest in size?
    a) single-phase
    b) two-phase
    c) three-phase
2140. In electricity, the current that flows in one continuous direction is:
    a) AC
    b) DC
    c) commutator

2141. The thickness of a fire bed with hand firing is:
   a) over 6"
   b) 2" to 6"
   c) 2" even
2142. The filter bed of open heaters is:
   a) coke
   b) stones
   c) gravel
2143. A common slide valve has:
   a) 2 lap edges
   b) 3 lap edges
   c) 4 lap edges
2144. To increase horsepower of a Corliss engine, you would never:
   a) adjust governor
   b) adjust valve travel
   c) decrease back pressure
2145. Tension relief on a single-element regulator is to:
   a) keep tension on bronze pivots
   b) relieve pressure produced by adjusting nut
   c) avoid excessive pressure on valve when boiler is cold
2146. The recommended way to start a centrifugal pump with no head is:
   a) suction and discharge valves closed
   b) suction open and discharge closed
   c) suction closed and discharge open
2147. Silica will carry over with steam at:
   a) 250 psi
   b) 350 psi
   c) 450 psi
2148. Steam would have difficulty separating from water at:
   a) 1000 psi
   b) 3200 psi
   c) 4500 psi
2149. The best coal to use in a traveling grate stoker is:
   a) the lower the ash content the better the coal
   b) coal should have 3% ash content
   c) coal of 7% ash or more should be used
2150. Water added to a steaming boiler through an economizer cannot be less than:
   a) 70°F
   b) 120°F
   c) 212°F
2151. The best torque in starting a motor is:
   a) compound
   b) series
   c) shunt

2152. Across the moving blades of a reaction turbine there is:
   a) a perceptible drop in pressure
   b) no perceptible drop in pressure
   c) an increase in absolute velocity
2153. A diffuser pump is:
   a) usually a centrifugal pump
   b) always a centrifugal pump
   c) a rotary pump
2154. Water from a sodium zeolite system will:
   a) increase dissolved solids
   b) decrease dissolved solids
   c) have the same amount of dissolved solids, as it is only on ion exchanger
2155. An interpole is used:
   a) on a DC generator
   b) on an AC generator
   c) for phasing
2156. Phasing is dependent upon:
   a) the number of poles
   b) not dependent on the number of poles
   c) the speed of the generator
2157. Speed of piston travel on head end to mid-point in relation to crank end is:
   a) faster
   b) slower
   c) the same
2158. Water circulation in a boiler tends to increase with the:
   a) increase in pressure
   b) decrease in pressure
   c) increase in load
2159. A crossbox used as a mud drum is found:
   a) on a box header
   b) on a sinuous header
   c) where water is particularly muddy
2160. More primary air would be used in a pulverizer with:
   a) semi-bituminous coal
   b) bituminous coal
   c) sub-bituminous coal
2161. You would correct a pulverizer fire by:
   a) using water
   b) increasing the fuel-to-air ratio
   c) using CO_2
2162. You could tell a pulverizer fire by the:
   a) outlet fuel-air temperature
   b) smoke at air tempering damper

2163. The difference between a traveling grate and a chain grate is:
  a) a chain grate is a traveling chain with bars, while traveling grate is not
  b) a traveling grate is a traveling chain with bars, while a chain grate has chain coming from above
  c) a chain grate is made of tiny chain bars while both are endless chains
2164. Steam is at its highest temperature at:
  a) 210 psi
  b) 450 psi
  c) 3200 psi
2165. Cushion valves, as used on a duplex pump, are on the:
  a) steam side
  b) water side
  c) both sides
2166. A shaft governor will assume its new position within:
  a) 16 revolutions
  b) 32 revolutions
  c) 64 or more revolutions
2167. The temperature the air can be heated to in a stoker furnace depends upon the:
  a) type of fuel being burned
  b) combustion rates
  c) steel castings
2168. A motor with constant speed as load is applied is a:
  a) shunt
  b) series
  c) compound
2169. In a synchronous motor, strength of DC affects the:
  a) speed of motor
  b) rotation of shaft
  c) power factor
2170. To burn one lb of typical fuel oil, the theoretical amount of air needed is:
  a) 12 to 14 lbs
  b) 14 to 16 lbs
  c) 16 to 18 lbs
2171. The amount of dissolved solids cannot be reduced by:
  a) removing carbonates
  b) removing sulphates
  c) running water through a pressure filter
2172. You will increase chlorides in a boiler if you:
  a) increase the pressure
  b) decrease the steaming rate
  c) run water to the boiler from a zeolite tank without flushing
2173. When steam is throttled in a reducing station from 200 psi to 15 psi:
  a) a higher temperature and superheat are reached
  b) the temperature of steam is increased and the steam is superheated

c) the steam temperature is increased and some of the steam is condensed
d) the steam will have a higher temperature than that corresponding to pressure, and the steam is superheated

2174. In using a feedwater test for chlorides, you would:
    a) take a fresh sample and neutralize it
    b) take a fresh sample and not neutralize it
    c) take the test after the P & M test

2175. If using fuel with a low fusing point, you would probably have:
    a) water screen
    b) dry bottom
    c) wet bottom

2176. A condenser that would not have injection water is a:
    a) barometric
    b) jet
    c) surface

2177. A centrifugal pump should:
    a) take its water 3' or more above heater
    b) take its water from the surface of the heater
    c) have vapor free water

2178. A device that has nothing to do with thrust is a:
    a) Kingsbury bearing
    b) shrouding
    c) dummy piston

2179. A constant head chamber is used on a:
    a) pressure regulator to control water level
    b) device to determine true water level in a boiler
    c) blow-down tank to insure full head of water

2180. When laying-up a boiler dry, the chemical to use in pans is:
    a) slaked lime
    b) unslaked lime
    c) sodium sulphite

2181. The ASME symbol on an HRT is:
    a) on the side of the boiler near the front head
    b) on the front head above the center line
    c) any place on the side in a conspicuous place

2182. An open feedwater heater can be used as a de-aerator because:
    a) an open heater can heat water to 210°
    b) gases won't stay in solution when heated near boiling point
    c) an open heater is open to the atmosphere

2183. Boiler horsepower can be figured on the:
    a) pressure and temperature of the boiler
    b) heating area only
    c) heat absorbing capacity of the boiler

2184. Heat recovery of 20 in feedwater would be equal to an efficiency of:
   a) 1%
   b) 2%
   c) 3%
2185. Factor of safety is:
   a) working pressure times 5
   b) ratio between bursting pressure and safe working pressure
   c) ratio between working pressure and bursting pressure
2186. What stoker has the largest combustion space for pounds of fuel burned?
   a) spreader
   b) underfeed
   c) traveling grate
2187. A slide valve engine with indirect valve and indirect valve gear, the eccentric:
   a) leads crank 90° plus angle of advance
   b) behind crank 90° plus angle of advance
   c) behind crank 90° minus angle of advance
2188. Convection is the principal mode of heat transfer in:
   a) radiation
   b) fluids
   c) gases
2189. By-passes should be installed on all valves over:
   a) 5"
   b) 8"
   c) 12"
2190. In waterwall circulation with headers below the mud drum, the take-off to the water-tube header is usually taken from:
   a) a water and steam drum
   b) the mud drum
   c) another header attached to the waterwall header
2191. To get the best turbulence in a pulverized fuel furnace, use:
   a) tangential firing
   b) horizontal firing
   c) vertical firing
2192. An 8" pipe would be measured by:
   a) nominal inside diameter
   b) nominal outside diameter
   c) outside diameter minus the thickness
2193. The area of the holes in a dry pipe for efficient operation have to be:
   a) larger than the area of the discharge pipe
   b) equal to the area of the discharge pipe
   c) twice the area of the discharge pipe
2194. At 150 psi, the temperature is:
   a) 358°

b) 330°
c) 256°

2195. Stacks should be protected from lightening by:
   a) a proper ground
   b) using a lightening rod
   c) design of stack

2196. The amount of friction of a well-lubricated reciprocating engine should be limited to:
   a) 10%
   b) 20%
   c) 30%

2197. If you were to tighten up on the adjusting nut of a single-element feedwater regulator, you would:
   a) raise the water level
   b) lower the water level
   c) cause the regulator to operate faster

2198. The safe working pressure of a bent-tube boiler is figured on the:
   a) ligament of the tubes
   b) longitudinal seam
   c) girth seam

2199. The common cause of bagging of the tubes is:
   a) not enough water circulation
   b) overheating of the tubes
   c) scale on the inside of the tubes

2200. If a staybolt were drilled 1/2" past the inside plate, the length of the staybolt would be:
   a) over 8"
   b) under 8"
   c) could be used on fire-tube boilers only

2201. Waste heat boilers, as compared to direct fired boilers, have:
   a) less clean outdoors
   b) more clean outdoors
   c) no clean outdoors

2202. Vertical water-tube boilers are supported by:
   a) legs on lower drum and allowed to expand upward
   b) hangers on upper drum and allowed to expand upward
   c) hangers on upper drum and allowed to expand downward

2203. A shaft bearing, when renewed, should be fitted by:
   a) scraping
   b) a machine fit
   c) filing

2204. Velocity is increased by increasing the speed of a centrifugal pump:
   a) at the square of the pump speed
   b) at the cube of the pump speed

c) directly proportional to the pump speed
2205. A major adjustment on a safety valve with two adjusting rings is made by:
   a) top ring only
   b) bottom ring only
   c) both rings
2206. In a Ringleman chart, maximum density is at:
   a) one
   b) five
   c) ten
   d) four
2207. To test for CO_2 in boiler feedwater, use:
   a) an orsat
   b) a CO_2 recorder
   c) a pH indicator
2208. For most efficient operation, pressure in open heaters is regulated by:
   a) regulating the vent valve
   b) throttling steam from the engine
   c) throttling steam from auxiliaries
2209. With adjusting ring attached to the disc of a safety valve by threads, in order to be sure the disc does not rotate when adjusting blow-back, adjust when:
   a) valve is detached from boiler
   b) the boiler is near popping
   c) there is no pressure on boiler
2210. The adjusting ring, when attached to valve seat by threads, when turned to the right:
   a) increases blow-back
   b) decreases blow-back
   c) decreases popping pressure
2211. A superheater is kept cool by:
   a) high steam velocity
   b) steam in contact with tubes
   c) both of the above
2212. With an increase in load, a convection superheater's temperature will:
   a) increase
   b) decrease
   c) have no effect
2213. With a decrease in load, a radiant superheater's temperature will:
   a) increase
   b) decrease
   c) remain the same
2214. The fine spray mist in an atomizing heater is created by high velocity:
   a) steam
   b) steam and water
   c) water coming into contact with heating steam

2215. Adding weight to the balls of a flyball governor:
　　a) increases speed of governor
　　b) decreases speed of governor
　　c) causes no change in speed of governor
2216. Priming can be noticed by:
　　a) water hammer
　　b) vibration of machinery
　　c) condensation in steam lines
2217. In a stoker-fired boiler, pressure in combustion chamber is:
　　a) below atmospheric
　　b) above atmospheric
　　c) balanced draft
2218. If a given weight of steam is superheated, the volume:
　　a) increases
　　b) decreases
　　c) remains the same
2219. One gallon of #6 fuel oil weighs:
　　a) 7 to 7½ lbs
　　b) 7½ to 8 lbs
　　c) 8 to 9 lbs
2220. The primary air in a pulverized-firing furnace is used for:
　　a) mill drying
　　b) carrying the fuel
　　c) increasing combustion rate
2221. Mean effective pressure of indicator card is measured from:
　　a) atmospheric pressure line
　　b) absolute pressure line
　　c) exhaust pressure line
2222. An altitude meter used in heating systems is graduated in:
　　a) pounds
　　b) inches of mercury
　　c) feet
2223. On an underfeed stoker, excess air is:
　　a) 25%
　　b) 35%
　　c) 45%
2224. Which flanges must be tightened the hardest?
　　a) smooth surface
　　b) rough surface
　　c) serrated surface
2225. In a properly fitted bearing, the bearing touches:
　　a) 0% of journal
　　b) 80% of journal
　　c) 100% of journal

2226. A boiler may support what percent of steam piping?
   a) any amount
   b) less than half
   c) none
2227. Water can contain the most dissolved solids at:
   a) high temperature
   b) low temperature
   c) temperature has no effect
2228. Which do you resurface first?
   a) valve
   b) valve seat
   c) either one
2229. In an air heater, as velocity of the flue gases increases, gas will:
   a) increase in temperature
   b) decrease in temperature
   c) have little effect
2230. A tandem compound engine has:
   a) two pistons side by side
   b) two pistons and one piston rod
2231. In a furnace that has waterwalls, there is:
   a) no danger of incomplete combustion through cooling
   b) danger of incomplete combustion through cooling
   c) incomplete combustion if forced
2232. Fire doors on a water-tube boiler should be:
   a) self-locking
   b) friction contact
   c) spring closed
2233. Most scale deposits will be found:
   a) on the water inlet
   b) in the rear bank of tubes
   c) in the tubes over the fire
2234. The thrust in a single suction centrifugal pump is to the:
   a) suction
   b) discharge
   c) is a radial thrust
2235. To change the direction of rotation of an induction motor (polyphase):
   a) change armature connections
   b) change the phases
   c) change any two AC power leads
2236. As steam pressure goes up, the weight of water an injector will handle:
   a) increases
   b) decreases
   c) remains the same

2237. If burning oil, and a change is made to pulverized fuel, furnace will need:
   a) a larger volume
   b) less volume
   c) no change is needed
2238. The advantage of a compound engine over a simple engine is:
   a) less moving parts
   b) less steam consumption
   c) takes up less space
   d) less condensation
2239. The best time to blow-down a boiler with waterwalls is:
   a) when it is off the line
   b) when it is steaming lightly
   c) during heavy loads
2240. The motor of a centrifugal pump has greatest load when valve is:
   a) throttled
   b) open
   c) closed
2241. Turbines run most efficiently at:
   a) high speed
   b) low speed
   c) variable speed with the load
2242. API degrees of #6 oil are:
   a) 7-22
   b) 22-32
   c) 32-42
2243. Temperature of bearing oil on a bearing is:
   a) 120°F
   b) 150°F
   c) 180°F
2244. The DC motor with the most starting torque is:
   a) series
   b) compound
   c) shunt
2245. A Jones stoker is:
   a) side feed
   b) underfeed
   c) overfeed
2246. The condenser that does not need a condenser pump is a:
   a) surface
   b) barometric
   c) jet
2247. Synchronous speed for two pairs poles 60 cycles is:
   a) 1200 rpm

b) 1800 rpm
c) 3600 rpm
2248. For equal horsepower, a fire-tube, compared to a water-tube boiler, has:
   a) more water space
   b) less water space
   c) same water space
2249. The increase of efficiency with air heaters is greatest with:
   a) spreader stoker
   b) pulverized coal
   c) underfeed stoker
2250. The use of air is necessary with:
   a) pulverized coal
   b) spreader stoker
   c) Detroit Roto-Stoker
2251. In thrust bearings, lubrication is due to:
   a) oil bath
   b) oil pressure
   c) wedge principle
2252. A piston valve engine with inside admission and direct valve gear, at release:
   a) valve and piston are traveling in same direction
   b) valve and piston are traveling in opposite directions
   c) inside admission has no relation to piston travel
2253. As compared to a simple engine, a compound engine requires a flywheel size that is:
   a) larger
   b) smaller
   c) the same size
2254. Cast iron flanges and nozzles on boilers of 100 psi:
   a) may be used
   b) may not be used
   c) can be used if not exposed to fire
2255. To reverse a DC motor, reverse:
   a) armature leads
   b) number of poles
   c) phases
   d) any two outside leads
2256. A non-releasing Corliss engine has a:
   a) shaft governor
   b) flyball governor
   c) pendulum governor
2257. A regenerative air preheater air duct is:
   a) the same size as gas duct
   b) smaller than gas duct
   c) larger than gas duct

2258. In a generator with 4 poles and 1800 rpm, the frequency is:
   a) 90 cycles
   b) 120 cycles
   c) 60 cycles
2259. The minimum size of a manhole plate is:
   a) 11 x 15
   b) 10 x 15
   c) 11 x 16
2260. The maximum pressure to which cast iron can be used on a boiler is:
   a) 160 psi
   b) 250 psi
   c) 450 psi
2261. The type of coal that is plastic is:
   a) coking
   b) caking
   c) volatile bituminous
2262. Coal that requires greater pulverization is:
   a) anthracite
   b) bituminous
   c) lignite
2263. Coal that would produce the most smoke is:
   a) semi-bituminous
   b) bituminous
   c) sub-bituminous
2264. Operation of an economizer is the most hazardous at a:
   a) light load
   b) heavy load
   c) banked load
2265. On a Copes feedwater regulator, the steam line from the boiler pitches:
   a) to the boiler
   b) to the regulator
   c) is level
2266. If a series generator became motorized, it would:
   a) reverse direction
   b) the windings would burn up
   c) use more current
2267. On an underfeed stoker, the limiting factor of preheated air is:
   a) rate of firing
   b) type of metal castings
   c) fusion temperature of coal
2268. "P" reading is referred to as:
   a) sodium sulfate
   b) sodium carbonate
   c) sodium phosphate

2269. Nuts on a safety valve gauge should be tightened:
- a) with a wrench
- b) with a stillson wrench
- c) finger tight

2270. On a pulverizer, the fuel that needs the most primary air is:
- a) anthracite
- b) bituminous
- c) lignite

2271. To maintain level water and prevent priming, use a boiler with:
- a) large water space to heating surface
- b) large steam to water space
- c) large drum internals

2272. On a cross compound engine, later cut-off on a high-pressure cylinder:
- a) increases power on high-pressure cylinder
- b) decreases power on high-pressure cylinder
- c) decreases power on low-pressure cylinder

2273. A chemical in a boiler that causes corrosion in piping is:
- a) dissolved gases
- b) dissolved solids
- c) ammonia

2274. A turbine is protected from excess pressure by a:
- a) condenser
- b) governor
- c) relief valve

2275. Grate that burns most coal per hour per square foot is a:
- a) traveling grate
- b) single retort
- c) hand fired

2276. Highest temperatures are obtained in heating fuel oil if heater is located:
- a) in tank
- b) after pump
- c) before pump

2277. Use a manometer for draft because it:
- a) has no moving parts
- b) has no parts to foul up
- c) gives more accurate readings

2278. In an impulse turbine, pressure through moving blades:
- a) increases
- b) decreases
- c) is the same

2279. When pressure is put on a steam gauge, the gauge forms:
- a) an oval
- b) a circle
- c) a square

2280. A flat gauge glass will use:
   a) mica
   b) rubber gasket
   c) shellac for sealing
2281. In an impulse turbine, velocity through stationary blades:
   a) increases
   b) decreases
   c) remains the same
2282. A venturi meter measures:
   a) quantity of flow
   b) quality of flow
   c) steam density
2283. Which seam is the strongest?
   a) girth
   b) longitudinal
   c) head
2284. The weakest part of a welded bent-tube multi-drum boiler is:
   a) tube ligaments
   b) tubes
   c) seams
2285. It would be easier to get oil into a boiler by using a:
   a) jet condenser
   b) surface condenser
   c) closed feedwater heater
2286. An air compressor uses gaskets of:
   a) asbestos
   b) rubber
   c) air wick
2287. A reaction turbine uses:
   a) stationary blades only
   b) rotating blades only
   c) stationary and rotating blades
2288. The time required for cut-off on a Corliss engine depends upon the:
   a) speed of engine
   b) releasing gear
   c) positive pressure in dash-pots
2289. Stationary blades in a reaction turbine are mounted on the:
   a) rotor
   b) casing
   c) shaft
2290. Expansion of steam piping is figured on:
   a) inches/hundred feet
   b) thousandths of an inch/hundred feet
   c) feet/hundred feet of pipe

2291. The number to divide by to find grains per gallon is:
   a) 17.1
   b) 0.058
   c) 0.58
2292. Combustion of coal on underfeed stokers takes place in:
   a) retort
   b) top of fuel bed
   c) extension grates
2293. An underfeed stoker, fired properly, has:
   a) fire in retort
   b) coal in retort
   c) ashes in retort
2294. A steam connection to a water column should pitch:
   a) to the column
   b) to the boiler
   c) be level
2295. A connection between piston rod and connecting rod is:
   a) crankpin
   b) cross head
   c) top head
2296. A sight glass on a motor bearing can have pipe from:
   a) top of glass only
   b) top and bottom of glass
   c) bottom of glass only
2297. Phosphorous in boiler metal makes the steel:
   a) hot short
   b) cold short
   c) is beneficial
2298. In a dry bottom furnace, the water screen helps:
   a) the slag to break into smaller pieces
   b) protect refractory
   c) reduce temperature of the molten slag
2299. Air leakage through a boiler setting will lower:
   a) furnace temperature
   b) CO_2
   c) combustion rate
2300. When cutting in an open feedwater heater to service, open:
   a) steam inlet
   b) steam inlet and then drain
   c) drain and then steam inlet
2301. If eccentric is advanced on a "D" slide valve engine, cut-off will:
   a) occur earlier
   b) occur later
   c) not be affected

2302. On a turbine with carbon rings, the rings:
   a) rotate with the shaft
   b) do not rotate with the shaft
   c) rotate slowly with the shaft (like oil rings)
2303. A pump in a fuel-oil system is protected by a:
   a) fuel oil strainer
   b) relief valve
   c) safety head
2304. Water circulation in a water-tube boiler, compared to a fire-tube, is:
   a) less rapid
   b) more rapid
   c) the same
2305. An induction motor:
   a) is always a three-phase motor
   b) can be a single-phase motor
   c) cannot be reversed
2306. The viscosity of a good grade of mineral oil for a centrifugal pump is:
   a) 300 SSU at 70°F
   b) 500 SSU at 70°F
   c) 700 SSU at 70°F
2307. A new boiler should be started:
   a) slowly (to dry out the brickwork)
   b) fast (to get heat through quickly)
   c) fast (to prevent gas temperatures below the dew-point)
2308. If flue gas temperature increases, CO_2 will:
   a) increase
   b) decrease
   c) remain the same
2309. The weight on a Copes feedwater regulator aids in:
   a) opening the valve
   b) closing the valve
   c) stabilizing the regulator
2310. If the speed of an engine suddenly increases, the water level in the boiler will:
   a) increase
   b) decrease
   c) remain the same
2311. When starting multiple-retort underfeed stoker fired boilers equipped with superheaters:
   a) start a small fire in middle of grates
   b) start a small fire across full width of grates
   c) warm up furnace with oil or gas burners
2312. The absolute steam velocity across the moving blades of reaction:
   a) increases

b) decreases
   c) does not change
2313. A marking not found on a safety valve is the:
   a) ASME symbol
   b) popping pressure
   c) capacity
   d) discharge pipe diameter
2314. If a fire occurs in an operating pulverizer:
   a) stop the fuel feed to the pulverizer
   b) increase the hot air flow to the pulverizer
   c) increase the raw fuel feed to the pulverizer
2315. The use of induction motors in a plant will:
   a) be an aid to the power factor
   b) not be an aid to the power factor
   c) not affect the power factor
2316. A mercury column reading of 2" of mercury is approximately equal to a pressure of:
   a) .5 psi
   b) 1 psi
   c) 2 psi
2317. A cubic foot of water weighs 62.4 lbs at 39°F. At 212°F it weighs:
   a) 59.8 lbs
   b) 61.0 lbs
   c) 62.4 lbs
2318. Waterwalls on a bent-tube boiler are fed from the:
   a) steam and water drum
   b) separate headers
   c) mud drum
2319. The reactionary force of steam escaping from the huddling chamber:
   a) increases blow-back
   b) decreases blow-back
   c) increases popping pressure
2320. A plain babbitt bearing should use an oil of:
   a) 300 SSU at 70°F
   b) 500 SSU at 70°F
   c) 700 SSU at 70°F
2321. The fire has been killed on a boiler that has low water. The first thing you would do is:
   a) close the steam stop valve
   b) speed up the boiler feed pump
   c) stop the boiler feed pump
   d) let the pressure drop
2322. Water-tube boiler drums are constructed:
   a) single-course

b) two-course
c) three-course
2323. If the combustion rate of fuel is doubled:
 a) combustion space must be doubled
 b) the rate of heat must be doubled
 c) the steam temperature is doubled
2324. One boiler horsepower, according to most local ordinances, is:
 a) the evaporation of 34.5 lbs of water from and at 212°F
 b) 33,475 Btu/hr
 c) 10 ft^2 of heating surface
2325. If water went out of sight on a steaming boiler, after pulling the fire, the next step would be to:
 a) increase pump speed
 b) wait for steam pressure to drop
 c) close steam stop valve
2326. The greatest temperature difference is between:
 a) gas film and tube
 b) tube and water and steam film
 c) water film and tube film
2327. To detect trap operation, you would:
 a) listen with a stethoscope
 b) feel the trap
 c) use a contact thermocouple
2328. An HRT boiler:
 a) is set level
 b) slopes from front to rear
 c) slopes from rear to front
2329. Thermosiphon arch in locomotive boilers:
 a) decreases boiler capacity
 b) decreases water circulation
 c) increases water circulation
2330. Lantern rings can seal:
 a) inward
 b) outward
 c) both ways
2331. The gravel in a zeolite softener is:
 a) above the bed
 b) below the bed
 c) above and below the bed
2332. The head that a centrifugal pump can pump against is:
 a) greater for water than for oil
 b) greater for oil than for water
 c) the same for oil and water

2333. If the inside wedges on the crank end of a connecting rod are taken up, the piston will:
   a) move forward
   b) move toward crank end
   c) not change position
2334. Lantern rings are used for:
   a) oil rings
   b) packing
   c) leakoff cavities
2335. Torching in recuperative air preheaters is caused from:
   a) fouling of the flue gas surfaces
   b) burning of the metal in the preheater
   c) a hole in the plate between flue gas and air passages
2336. To change the voltage on an induction motor if it is a squirrel cage motor:
   a) change the stator
   b) change the rotor
   c) both a and b
2337. A generator with a power factor of 80 is most efficient at:
   a) 80% power factor
   b) 100% power factor
   c) 60% power factor
2338. A resistance thermometer operates on what principle?
   a) ohmmeter
   b) resistance of coil of nickel and platinum compared to standard
   c) resistance of two dissimilar wires due to voltage generated causes temperature rise
2339. What advantage has a compound engine over a simple engine?
   a) less moving parts
   b) less steam consumption
   c) takes up less space
2340. What motor has the most torque?
   a) series
   b) compound
   c) shunt
2341. Btu necessary to generate one kilowatt hour (kWh):
   a) 1,000
   b) 2,000
   c) 3,000
2342. What effect would increasing of phases have on frequency of motor?
   a) increase
   b) decrease
   c) no effect
2343. Steam escaping from a safety valve tends to:
   a) increase blow-back

b) decrease blow-back
c) not change blow-back

2344. Which has the most drop through an air preheater?
   a) gas
   b) air
   c) both

2345. Which will give lowest heat by convection?
   a) natural gas
   b) oil
   c) coal

2346. Which will give lowest heat by radiation?
   a) natural gas
   b) oil
   c) coal

2347. In taking a test with an orsat:
   a) .002 is the gas analyzed first
   b) 2 is the first found
   c) it would not matter which was taken off first

2348. In taking a test with an orsat:
   a) the water in the bottle must be pure distilled water
   b) plain tap water is sufficient
   c) any fluid could be used

2349. The terms "caking coal" and "coking coal":
   a) mean the same thing when applied to coal
   b) refer to different qualities of coal
   c) refer to type of ash left after burning coal

2350. Why is smoke-free combustion so important?
   a) the presence of smoke indicates that combustion has not been complete and therefore fuel has been wasted
   b) it contaminates air and makes neighborhood look bad
   c) it violates the smoke abatement code

2351. The Hargrove method is one of two distinct methods for:
   a) determining the grindability index of coal
   b) determining fusion temperature of coal
   c) analyzing flue gas

2352. It is necessary to pulverize fuel so it will pass through 200 mesh:
   a) in order that ignition is rapid and sure
   b) because larger size particles would "plus up" burner
   c) to decrease size of furnace

2353. Most pulverizer specifications require an output of not more than 2% plus 50 mesh fuel because:
   a) more than 2% plus 50 mesh causes excessive slagging in furnace
   b) more would cause excessive wear of burner
   c) burner would plug up if larger sizes were used

2354. When pulverizing coal:
   a) low-volatile coal should be pulverized to a higher degree of fineness than high-volatile coals
   b) high-volatile coal should be pulverized the finest
   c) does not make any difference
2355. In most cases, gaseous fuels are measured and purchased in terms of cubic feet. In some cases they are purchased by heating values expressed as "therms". One therm is equal to:
   a) 100,000 Btu
   b) 75,000 Btu
   c) 50,000 Btu
2356. Manufactured gas (such as coal gas or coke over gas) usually has a heating value of:
   a) 500 to 600 Btu/ft^3
   b) 700 to 900 Btu/ft^3
   c) 900 to 1100 Btu/ft^3
2357. Gas burners should never be lighted until:
   a) furnace has been purged at least 5 minutes at an air flow equal to at least 50% of full load
   b) furnace has been purged at least 2 minutes, being sure back damper is open
   c) it is not necessary to purge boiler if it has not been fired recently
2358. All things being equal, you could check viscosity of oil more quickly by using:
   a) Saybolt Universal Viscosimeter
   b) Saybolt Furel Viscosimeter
   c) there wouldn't be too much difference in time required
2359. The reason for wide variation of heating values of various gases is due to:
   a) the percent of hydrocarbons present
   b) the amount of oxygen (O_2) present
   c) the difference in the weight of various gases
2360. When burning gas, the best type of flame is a:
   a) short, non-luminous flame
   b) pulsating flame
   c) long, pale blue flame that is tipped with yellow
2361. The pressure on high-pressure gas burners used in large furnace installations is from:
   a) 20 to 30 psi
   b) 40 to 50 psi
   c) 60 to 100 psi
2362. The common term used for so-called tangential firing is:
   a) corner firing
   b) angle firing
   c) mixture firing

2363. Modern, large-capacity units are built with:
   a) one upper drum to lower cost
   b) sometimes two drums to provide sufficient steam-relieving capacity
   c) three drums in order to be able to install tubes at a 90° radius to drums
2364. The most important element in determining dew point is:
   a) absolute humidity of the air
   b) relative humidity of the air
   c) sulphur content of the coal
2365. In a non-steaming economizer, the temperature of water should be raised to:
   a) 34 to 45 degrees of saturation temperature
   b) 45 to 55 degrees of saturation temperature
   c) 55 to 65 degrees of saturation temperature
2366. A mollier chart is used for determining:
   a) enthalpy
   b) entropy
   c) steam conditions
2367. When taking an indicator card, the first event taken is:
   a) atmospheric line is drawn
   b) card is taken
   c) indicator should be warmed up
2368. What is meant by clearance on an engine?
   a) space between cross head and cylinder when at head end dead center
   b) space between piston and cylinder with piston at head end dead center
   c) space between piston and cylinder plus steam passages with piston on dead center
2369. What is the theoretical efficiency of an engine?
   a) ratio of heat rejected to heat received
   b) ratio of heat received to heat rejected
   c) ratio of heat extracted to heat received
2370. Define the term "adiabatically":
   a) steam expanded to greater volume than normal for given pressure
   b) steam compressed to greater pressure than input pressure
   c) steam, when expanded or compressed, does not lose or gain any heat by heat transfer
2371. What is a condensing engine?
   a) an engine which operates with back pressure less than atmosphere
   b) an engine which operates with back pressure equal to or greater than atmospheric pressure
   c) an engine which condenses steam before exhausting
2372. At what speed does a low-speed engine operate?
   a) 200 rpm or less
   b) 110 to 200 rpm
   c) 80 rpm or less
   d) 40 rpm or less

2373. At what throttle pressure does a low-pressure engine operate?
   a) 150 psi or more
   b) 80 to 150 psi
   c) 80 psi or less
   d) 40 psi or less
2374. How does a shaft governor control speed?
   a) throttles input steam
   b) controls valve action
   c) opposes speed by opposing direction of rotation of flywheel
2375. How does a flyball governor control speed?
   a) throttles input steam
   b) controls valve action
   c) changes pressure of steam
2376. What is an indicator diagram used for?
   a) analysis of what is taking place in engine cylinder
   b) indicates how much water is entering with steam
   c) indicates the speed of the engine
2377. What is meant by a double-acting engine?
   a) an engine with two pistons
   b) an engine which takes steam on both strokes
   c) an engine which takes steam on one stroke with flywheel momentum carrying engine through return stroke
2378. Steam is used extensively to:
   a) increase the ratio of work done to heat used
   b) decrease the amount of back-pressure
   c) decrease the size of cylinder needed for given work output
2379. When an engine is running over, it means that:
   a) oil is running over
   b) the engine is running over the speed designed for
   c) the top of the flywheel is turning away from the cylinder
2380. An intercooler is used with:
   a) an open heater
   b) a closed heater
   c) an air compressor
2381. Which operates first on an element Copes feedwater regulator?
   a) thermostatic tube
   b) diaphragm
   c) orifice
2382. At what temperature does silica carry over with steam?
   a) 650°
   b) 250°
   c) 450°
2383. The mechanical equivalent of heat by the US Bureau of Standards is:
   a) 10 ft^2 heating surface

b) the number of Btu/minute or Btu/hour
c) the number of foot-lbs in a Btu

2384. How many Btu in one horsepower minute?
 a) 33,000 Btu
 b) 2,545 Btu
 c) 42.42 Btu

2385. The range of which thermometers are made would be capable of indicating temperatures ranging from:
 a) 150 to 1000°F
 b) 20 to 3000°F
 c) -10 to 212°F

2386. For measuring extremely high or low temperature, use a:
 a) hygrometer
 b) pyrometer

2387. At what temperature does water, at maximum density, weigh more per unit of volume?
 a) 32°F
 b) 63°F
 c) 39°F

2388. For a given pressure, the temperature and density of a saturated vapor are:
 a) fixed
 b) vapor is a few degrees higher
 c) the temperature and density of the vapor have no relation

2389. In boiler operation, the furnace heat is transferred by:
 a) conduction
 b) radiation only
 c) convection and radiation

2390. A soot blower connection, attached to the outlet from the superheater or reheater that is used for safety valve connection:
 a) is not allowed by code
 b) a soot blower connection may be attached

2391. Expansion and contraction of steel rolled soft per lineal inch 1 F.A.H. increases or decreases:
 a) .000005
 b) .00005
 c) .005

2392. Practically all of the earth's heat is derived, directly or indirectly, from the:
 a) sun
 b) volatile matter under the earth's surface
 c) trade winds from constant temperature of oceans

2393. Large condensing steam engines use what percent less steam than non-condensing engines?
 a) 10% to 20%

b) 20% to 40%
c) 40% to 50%

2394. Steam engines converted from non-condensing to condensing operation will develop:
  a) less horsepower
  b) more horsepower
  c) same horsepower

2395. Slide valves are employed in engines where:
  a) simplicity and low price are more important
  b) steam consumption needs to be kept down
  c) economy of the engine is important

2396. Proper load for any slide valve should be:
  a) determined by an indicator
  b) 3¼" for each foot of stroke
  c) determined by quiet operation

2397. The average life of a low-speed steam engine is:
  a) 17 years
  b) 20 years
  c) 28 years

2398. Economy resulting from a feedwater heater using exhaust steam is:
  a) 5 to 10%
  b) 11 to 14%
  c) 14 to 20%

2399. A primary or vacuum heater is located:
  a) between engine and condenser
  b) between feed pump and boiler
  c) location varies

2400. A ring valve is used on:
  a) injectors
  b) simplex pumps
  c) steam engines

2401. Tempering of coal means:
  a) amount of moisture it contains as fired
  b) amount of moisture when found
  c) amount of carbon

2402. Correct tempering of coal would be:
  a) 12% to 15% moisture
  b) less than 3% moisture
  c) none of the above

2403. Heavy fuel oil must be continually circulated past a burner:
  a) in order to return oil not burned to tank
  b) because temperature at burner would change with firing rate, thus a variation of oil viscosity would affect controls
  c) more carbon would form in furnace

2404. For equal horsepower, the HRT has:
   a) more water space than water tube
   b) less water space than water tube
   c) same water space as water tube
2405. Air is:
   a) an element
   b) a compound
   c) a mixture of elements
2406. The Hargrove method is a means of determining:
   a) grindability of coal
   b) boiler horsepower
   c) engine horsepower
2407. The Dulong formula has to do with:
   a) absolute temperature
   b) heat values of fuel
   c) heat balance of a plant
2408. The highest efficiency that can be economically justified in a boiler is about:
   a) 90%
   b) 80%
   c) 70%
2409. For a given set of steam conditions and firing methods, it is cheaper to build:
   a) one larger boiler
   b) two smaller boilers
   c) the cost is about the same
2410. A micron is:
   a) .25" to .00001"
   b) .1001" to .00004"
   c) 1"
2411. In burning fuel in suspension, the fuel is converted into a combustible by:
   a) the firing equipment
   b) the heat of the furnace
   c) the pulverizer
2412. Smoke and soot indicate incomplete combustion, which can be best controlled by:
   a) increasing excess air
   b) time, temperature and turbulence
   c) reducing fire rate
2413. The rotary pump is like the centrifugal pump in that they both:
   a) are positive displacement pumps
   b) give a non-pulsating flow
   c) have and use poppet valves
2414. The use of preheated feedwater has overcome:
   a) scale and corrosion

     b) the excess of expansion and contraction of boiler metals
     c) heat balance
2415. Can a condensate line dump to a condensate tank?
     a) high and low pressure steam traps cannot by used in the same line
     b) high or low pressure traps can dump in the same line
     c) a separate line must be used for high and low pressure traps
2416. With a full fluid film of oil between bearing surfaces:
     a) moves only with rotating bearings
     b) does not move at all
     c) none of the above
2417. In thrust bearings, lubrication due to floating pads is accomplished by the:
     a) wedge principle
     b) oil bath
     c) oil pressure
2418. In AC motors, speed is determined by:
     a) rheostat
     b) frequency of power and number of poles
     c) the number of poles, the faster the speed
2419. Cavitation of pump impellers is caused by:
     a) a positive suction head
     b) a negative suction head at or near vaporizing pressure
     c) too high a liquid temperature
2420. In impulse-type turbines, the diaphragm:
     a) is part of the rotor
     b) holds the nozzles and is a stationary part of the turbine
     c) holds rotating buckets
2421. What percent of steam does useful work on engines?
     a) 10%
     b) 25%
     c) 16%
2422. Brake horsepower is:
     a) more than indicated horsepower
     b) less than indicated horsepower
     c) the same as indicated horsepower
2423. The critical temperature and pressure of steam is:
     a) 3206 psi
     b) 3105 psi
     c) 4001 psi
2424. An indicator spring may be used indefinitely and must be:
     a) tested periodically
     b) replaced every 6 months
     c) replaced every year
2425. When mercury is used in a "U" tube draft gauge, meniscus is:
     a) convex

b) concave
c) level

2426. The bottom of meniscus is read when using:
   a) oil
   b) mercury
   c) water

2427. How many moving parts on a continuous flow steam trap?
   a) 3
   b) 1
   c) 0

2428. The top of meniscus is read when using:
   a) oil
   b) mercury
   c) water

2429. Safety valve on a closed feedwater heater should be set:
   a) 15 to 20 lbs above boiler pressure
   b) 15 to 20 lbs below boiler pressure
   c) the same at boiler pressure

2430. The average amount of water lost to evaporation when using spray pond is:
   a) 3%
   b) none
   c) 7%

2431. In a large industrial plant, an engine used just for power should be more efficient to operate:
   a) condensing
   b) non-condensing
   c) same

2432. A properly staged volute pump is in:
   a) radial balance
   b) axial balance
   c) radial and axial balance

2433. In the coal used in an average plant, the fusion point of ash is:
   a) 2100 degrees
   b) 3100 degrees
   c) 4100 degrees

2434. What is the ratio of heat absorbtion per square foot in an air preheater?
   a) 2
   b) 3
   c) 4

2435. A modern central-station boiler might contain:
   a) 1 mile of tubes
   b) 35 miles of tubes
   c) 70 miles of tubes

2436. A modern central-station boiler might contain:
   a) 0.75 lb
   b) 2.5 lbs
   c) 3.0 lbs
2437. Which pump has the most valves?
   a) rotary
   b) simplex
   c) duplex
2438. The common cause of bagging on tubes is:
   a) overheat of tube surfaces
   b) corrosion
   c) not enough water circulation
2439. The best place to install a throttling calormotor is:
   a) a straight length of pipe in which steam flow is descending
   b) a straight length of pipe in which steam flow is rising
   c) some pipe bond in horizontal length of pipe
2440. Economizers used in waste heat boilers, in relation to degree of temperature drop:
   a) are more efficient than direct-fired boilers
   b) are less efficient than direct-fired boilers
   c) have the same effect
2441. The conductivity motor is used to determine:
   a) the amount of ionizing boiler water salts
   b) measure of suspended matter such as organic material and un-ionized solids
   c) the amount of hardness is ppm
2442. For the same fuel burning rate, if the temperature of feedwater was lowered:
   a) the degree of superheat steam would increase
   b) the temperature of superheat is not affected by the temperature of feedwater
   c) the degree of superheat steam would decrease
2443. Gas weights and temperatures entering superheater of equal capacity will:
   a) vary approximately to the heat absorbed per lb. of steam
   b) not change
   c) change when feedwater temperature is lowered 50°
2444. The reheater or resuperheater must be installed to receive:
   a) radiant heat only
   b) convection heat only
   c) could be both radiant and convection heat
2445. The reheater will usually reheat superheated steam:
   a) too near the original temperature
   b) too near the original pressure
   c) it increases both the temperature and pressure too near original

2446. The average weight of bituminous coal is:
   a) 50 lbs
   b) 40 lbs
   c) 60 lbs
2447. The actual capacity of a new air compressor is:
   a) more than piston displacement
   b) less than piston displacement
   c) the same as piston displacement
2448. Mineral wool for insulation is made from:
   a) binet furnace slag
   b) vegetable and animal fibers
   c) cellular glass
2449. The best insulating material for steam temperature condition is:
   a) mineral wool
   b) asbestos
   c) plaster of Paris
2450. Insulating materials made of animal or vegetable fibers can only be used for:
   a) low temperatures
   b) high temperatures
   c) moderate temperatures
2451. From an economic standpoint:
   a) the thicker the insulation, the greater the savings
   b) the value of the heat saved does not increase in direct proportion to the thickness of insulation
   c) the weight of the secondary surface would be most important
2452. The loss per lineal foot per hour of bare steam pipe is:
   a) more with superheated steam
   b) more with saturated steam
   c) equal
2453. To acquire the ultimate maximum temperature required for the combustion of fuel oil, the fuel oil is heated:
   a) on suction side of pump
   b) on discharge side of pump
   c) under hood of tank
2454. The constant-speed DC motor having the most torque is:
   a) series
   b) shunt wound
   c) compound wound
2455. On a solid connecting rod with the inside wedges taking up wear would:
   a) decrease clearance volume
   b) lengthen rod
   c) shorten rod
2456. In a piston valve engine with inside admission and indirect valve, when the valve cuts off steam:

a) valve and piston are traveling in same direction
b) valve and piston are traveling in opposite direction
c) none of the above

2457. A compound engine with crank set 90° apart would:
a) need a receiver
b) exhaust directly to low-pressure cylinder
c) not need a receiver

2458. Manganous in boiler metal:
a) makes metal cold short
b) makes metal hot short
c) is considered desirable

2459. How many lbs of sulphite are required to absorb 1 lb of CO_2?
a) 10 lbs of sodium sulphate
b) 8 lbs of sodium sulphite
c) 5 lbs of calcium sulphate

2460. Steam that does not contain H_2O in any proportion is:
a) saturated steam
b) dry steam
c) wet steam

2461. The furnace sheet on a firebox is supported by:
a) stay bolts
b) diagonal stays
c) corrugated sheet

2462. Purification of steam and prevention of carry-over are achieved by:
a) dry pipe
b) steam
c) drum internals

2463. Heating water above the boiling point in an open heater with a continuous vent to atmosphere is considered inefficient because:
a) water flashes to steam
b) of loss of condensate to atmosphere
c) it causes excess pressure

2464. A flyball governor on a Corliss engine controls:
a) angle of advance
b) cut-off
c) admission

2465. Which gives the dirtiest steam?
a) submerged tubes
b) through tubes
c) none of the above

2466. The safety valve on a reheat cycle is:
a) needed because of boiler and superheater
b) needed to relieve total amount of steam going through reheater

c) needed, but only to be large enough to relieve 50% of saturated steam of boiler
2467. Which of these companies does not make boilers?
   a) Fairbanks Morse
   b) Murray
   c) Kooler
   d) Wickes
   e) Cliff
2468. On high-pressure boilers of large capacity:
   a) thickness of shell is same throughout
   b) shell is made thicker around tubes because tube ligament would be weak
2469. A bottle is:
   a) used to feed water to a burner
   b) a device used to slug feed chemicals to boiler
   c) a dry drum on a boiler
2470. What is ppm of hardness of typical river water?
   a) 50
   b) 100
   c) 256
2471. A regenerator air preheater cold duct, as compared to a gas duct:
   a) would be same size
   b) gas duct is larger
   c) air duct is larger
2472. An Adamson ring aids in:
   a) rigidity of boiler
   b) supports flues
   c) cleaning dirt from between boilers
2473. Retarders are in:
   a) fire-tube boiler tubes
   b) water-tube boiler tubes
   c) blow-down line to slow water to tank
2474. Thermal conductance is:
   a) Moht
   b) Thom
   c) Mode
2475. Bifurcated tube construction is used on:
   a) economizers
   b) waterwalls
2476. The frequency of a synchronous water 4 poles turning at 1800 rpm would be:
   a) 25 cycles
   b) 60 cycles
   c) 90 cycles
2477. In laying up a boiler, what chemical is used to control O_2?
   a) sodium sulphite

b) sodium sulphate
   c) sulphate
2478. What does a 600 hp boiler require by code?
   a) one electric pump
   b) one steam pump and one electric pump
   c) two electric pumps
2479. If fresh coal is added to a furnace, the most smoke by adding O_2 would be:
   a) near flame
   b) in retort
   c) nearer stack
2480. Lowest temperature water to put into an economizer is:
   a) 70°F
   b) 140°F
   c) 212°F
2481. Tri-sodium is:
   a) sometimes used to adjust pH
   b) never used
   c) used to inhibit caustic embrittlement
2482. Sodium sulphate and sodium carbonate ratio is maintained to:
   a) inhibit caustic embrittlement
   b) reduce scale
   c) reduce O_2
2483. Where would you find steam blanket?
   a) internal downcomers
   b) steam on top of water
   c) steam in risers
2484. What is constant head chamber used for?
   a) liquid level indicator
   b) height of water in blow-down tank
   c) height of water in water column
2485. When does a piston travel faster?
   a) first half of stroke
   b) second half of stroke
   c) same
2486. What condenser doesn't require supply pump once flow has started?
   a) barometer condenser
   b) surface condenser
   c) jet condenser
2487. The amounts being equal, which would remove more hardness?
   a) soda ash
   b) caustic soda
   c) Sanosite
2488. If a series generator became motorized, it would:
   a) reverse the rotation of the field

b) the field would remain the same
c) burn up windings
2489. Which causes most blow-down?
   a) sulphate
   b) carbonate
   c) silica
2490. If you had one generator driven by a turbine and one driven by a reciprocating engine, which would need more poles?
   a) reciprocating engine
   b) turbine engine
   c) they would need the same amount
2491. In a closed feedwater heater, which has more pressure?
   a) steam space
   b) water space
   c) same
2492. When tops of drums are at the same level on a Sterling boiler, circulation tubes are between:
   a) all drums
   b) 2 front drums
   c) 2 back drums
2493. Riser enters drum:
   a) below water level
   b) above water level
   c) both
2494. If the sulphur content of #6 oil increases, the Btu content will:
   a) decrease
   b) increase
   c) stay the same
2495. On a Corliss releasing engine, it is not considered good practice (when engine is running a maximum load) to increase power by:
   a) varying cut-off
   b) adjusting linkage to governor
   c) lowering exhaust pressure
2496. After a turbine is shut down, the auxilary most likely to be left running is:
   a) auxilary oil pump
   b) water to gland cooling
   c) air to injector
2497. On a nozzle-type safety valve, the operating principle is:
   a) huddling chamber principle
   b) reaction-type principle
2498. On a Copes water regulator, tension relief is:
   a) an aid in closing valve
   b) to reduce strain on linkage when boiler is down
   c) an aid in opening valve

2499. On an underfeed stoker, the factor that determines the temperature to which preheated air can be used is:
 a) rate of firing
 b) type of metal in stoker
 c) fusion temperature of coal
2500. Stays are used on which type of boiler?
 a) Wickes vertical
 b) Keeler CP boiler
 c) Wickes "A" type
2501. A "P" reading is usually referred to as the amount of:
 a) sodium sulphate in water
 b) sodium carbonate in boiler water
 c) sodium sulphite in boiler water
2502. On a centrifugal pump, if you should throttle discharge valve while the pump is running, you would:
 a) overload the driver
 b) unload the driver
 c) have no effect on the driver
2503. What is the speed of a regenerative-type air preheater?
 a) 7 to 10 rpm
 b) 2 to 3 rpm
 c) 10 to 25 rpm
2504. When the release takes place on an engine with a piston-type valve, the valve is moving in:
 a) same direction as piston
 b) opposite direction as piston
2505. What is the oil temperature coming from the bearing of a turbine?
 a) 120°F
 b) 140°F
 c) 180°F
2506. In a plate-type air heater, the curved vanes are on the:
 a) air side
 b) gas side
 c) both sides
2507. On a horizontal reciprocating machine, the direction of rotation is determined by the:
 a) AC or DC motor
 b) Mayors cut-off valve arrangement
 c) power, whether it is from or to the shaft
2508. A steam flow motor with a reservoir:
 a) prevents pulsation
 b) insures a full head of water
 c) insures a full head of oil

2509. On a forced-circulation boiler, what insures that there is water in all the tubes.
   a) sizing of tubes
   b) orifice in tubes
   c) pump size
2510. Bubble caps are used as:
   a) evaporators
   b) remote sight glass
   c) sediment at bottom of water chamber
2511. Riser tubes from waterwalls discharge steam and water:
   a) below water line
   b) above water line
   c) at water line
2512. The size and shape of recirculators are to:
   a) prevent carryover
   b) superheat steam
   c) equalize water level
2513. In a 4 drum Sterling boiler, water is:
   a) equal in all drums
   b) high in rear drums
   c) lower in rear drums
2514. Water velocity in risers, as compared to downcomers, is:
   a) greater in risers
   b) greater in downcomers
   c) the same in both
2515. A centrifugal pump with single suction is in hydraulic:
   a) axial balance
   b) radial balance
2516. A centrifugal pump with double suction is in hydraulic:
   a) axial balance
   b) radial balance
2517. An evaporator should have a:
   a) safety valve
   b) drain
2518. If you add bicarbonates to a boiler containing carbonates, you will have:
   a) biborato CO_2
   b) increased carbonates
   c) none of the above
2519. Surface blowoff should not exceed:
   a) 1"
   b) 1 ½'
   c) 2 ½'
2520. A 20" x 40" plate supporting 10 tons has:
   a) 250 psi

b) 25 psi
c) 15 psi
2521. The velocity of flue gas is:
   a) 4,000 to 6,000 fpm
   b) 6,000 to 8,000 fpm
   c) 8,000 to 10,000 fpm
2522. Expansion of water gauge glass is:
   a) .000643"
   b) .00034"
   c) .034"
2523. Safety valves, when mounted:
   a) the spindle must be vertical and be in an upright position
   b) can be located below water level provided steam line is taken from highest steam space
2524. Carbonate hardness is:
   a) 7
   b) sulphates
   c) bicarbonates of calcium and magnetism
2525. An impact tube means or measures:
   a) pressure
   b) velocity pressure
   c) total pressure
2526. Waterwell expansion is:
   a) downward
   b) upward
   c) either upward or downward
2527. When draining an economizer, temperature should be:
   a) above 200°F
   b) below 400°F
   c) below steam temperature
2528. When flue gas temperature gets too hot at preheater:
   a) bypass air around heater
   b) bypass gas around heater
   c) either a or b
2529. When two safety valves of different sizes are placed on the same line, the smaller must discharge:
   a) 50% of the larger
   b) 75% of the larger
   c) 25% of the larger
2530. To remove condensate from a condenser, use a:
   a) vacuum pump
   b) syphon
   c) steam trap

2531. The basic principle in adjusting the blow-back on a safety valve is:
   a) spring pressure
   b) valve disc area
   c) kinetic energy of steam
2532. An auto-transformer has:
   a) 1 winding
   b) 2 windings
   c) multi-windings
2533. An oil burner uses _____ steam.
   a) dry
   b) wet
   c) superheated
2534. On a compound engine, steam goes from high-pressure cylinder to:
   a) low-pressure cylinder and then to receiver
   b) receiver and then to low-pressure cylinder
   c) none of the above
2535. The primary purpose of not having alkalinity too high is to prevent:
   a) caustic embrittlement
   b) carryover
   c) none of the above
2536. When a return trap dumps, it requires:
   a) same volume of steam to water
   b) more volume of steam to water
   c) less volume of steam to water
2537. A furnace, when ash is abrasive or clinker, uses:
   a) chrome refractory fire brick
   b) glazed brick
   c) none of the above
2538. There may be a piping connection between the boiler and a water column:
   a) a piping damper regulator, on the steam gauge only
   b) no piping whatsoever
   c) piping that does not allow an appreciable amount of steam flow
2539. Reaction turbines are:
   a) velocity compounded
   b) pressure compounded
   c) none of the above
2540. With power at zero, the:
   a) voltage is zero
   b) current is zero
   c) both current and voltage are zero
2541. Chlorides are tested:
   a) before alkalinity
   b) after alkalinity
   c) before hardness

2542. Can cast iron flanges be connected to a boiler?
    a) may be connected
    b) only on boilers under 100 psig
    c) can never be used
2543. Double blow-down valves in one casting:
    a) are typically not permitted by local code
    b) may be used if the operation of one valve does not affect the other valve
    c) none of the above
2544. Water screens to cool the ash would be:
    a) tubes connected to mud drum
    b) water pumped separately
    c) in regular boiler circulation
2545. A Copes tension relief:
    a) prevents valve from hunting
    b) relieves valve from starting in cold start
    c) aids in thermostat expansion
2546. The perpendicular lines on an indicator card are taken from:
    a) atmospheric line
    b) exhaust line
    c) none of the above
2547. Zeolite softeners remove:
    a) temporary hardness
    b) permanent hardness
    c) some temporary and some permanent hardness
2548. Proportioning pump is used on:
    a) zeolite softener
    b) hot soda lime
    c) none of the above
2549. Weight of shaft on horizontal engines is:
    a) 80%
    b) 90%
    c) 100%
2550. Underfeed stoker use of preheated air is determined by:
    a) air temperature
    b) furnace temperature
    c) metal of grates
2551. A boiler of 250% efficiency has:
    a) 4 ft^2
    b) 6 ft^2
    c) 12 ft^2
    d) 8 ft^2
2552. Automatic gauge glasses are:
    a) round

b) flat
c) round and flat

2553. An injection pump is used with a:
   a) barometric condenser
   b) surface condenser
   c) low-level jet condenser

2554. Diffuser pumps are:
   a) always volute
   b) usually volute
   c) never volute

2555. A tandem compound engine has:
   a) a large flywheel
   b) the same rod for both pistons
   c) a receiver

2556. A regenerative air preheater gas duct is:
   a) smaller than air duct
   b) larger than air duct
   c) none of the above

2557. You would control CO_2 by:
   a) sodium sulphite
   b) sodium sulphate
   c) none of the above

2558. When laying up a boiler using the dry method, use:
   a) slaked lime
   b) unslaked lime
   c) none of the above

2559. In boiler water, the ratio of CO_2 to sulphite is:
   a) the same
   b) 8 lbs sulphite to 1 lb of CO_2
   c) none of the above

2560. A 600 hp boiler needs:
   a) a steam and electric pump (coal-fired)
   b) a steam pump only
   c) an electric pump only (gas, oil, pulverized)

2561. A water column for 350 psi is made of:
   a) steel
   b) cast iron
   c) none of the above

2562. On a modern high pressure boiler:
   a) the solid refractory walls have the thickest insulation
   b) the air-cooled walls don't need insulation
   c) the air-cooled walls have the thickest insulation

2563. To raise alkalinity, it is best to use:
   a) monosodium phosphate

b) disodium phosphate
c) trisodium phosphate
2564. Plate-type air preheated baffles are used on the:
   a) gas side
   b) air side
   c) none of the above
2565. Fire brick refractory for high temperature is:
   a) silicon carbide
   b) plastic
   c) tile brick
2566. Steam at 100 psi reduced to 15 psi would be:
   a) wet
   b) saturated
   c) superheated
2567. Less discharge on a centrifugal pump with:
   a) oil
   b) water
   c) same
2568. Lower discharge ring on a safety valve:
   a) decreases blow-back
   b) increases blow-back
   c) same
2569. A compound engine:
   a) needs a receiver
   b) does not need a receiver
   c) none of the above
2570. Tubes are measured in:
   a) ID
   b) OD
   c) monical
2571. A multi-stage centrifugal pump is:
   a) volute
   b) diffuser
   c) none of the above
2572. Tighten the adjusting nut on a Copes to:
   a) raise the water in the boiler
   b) lower the water in the boiler
   c) none of the above
2573. Strength of DC on a synchronous motor will affect:
   a) speed
   b) power factor
   c) none of the above
2574. A gasket suitable for condensate return is:
   a) cork

b) neoprene
c) asbestos
d) rubber

2575. Do not wipe oil tank or piping with:
a) sponges
b) rags
c) cotton waste

2576. What percent of the stean piping is supported by the boiler?
a) 10%
b) 0%
c) 20%

2577. A chimney produces draft by:
a) radiation
b) conduction
c) convection

2578. Boiler fired underfeed stoker changed to oil would:
a) decrease capacity
b) increase capacity
c) none of the above

2579. Which given furnace will have greater capacity?
a) coal
b) gas
c) oil

2580. Viscosity of oil will decrease:
a) when heated
b) to 250° and then level off
c) none of the above

2581. It is best to heat oil with:
a) wet steam
b) dry steam
c) superheated steam

2582. A governor of a small turbine will change speed:
a) less than 1°
b) up to 2°
c) 10°

2583. A shaft governor works on:
a) centrifugal force
b) inertia
c) both

2584. To warm up a water tank on an open feedwater heater:
a) run steam through vent
b) throttle steam to warm up slowly
c) open quickly

2585. To increase horsepower of a Corliss engine, you would never:
 a) adjust governor
 b) adjust valve travel
 c) decrease back pressure
2586. Tension relief on a single-element regulator is to:
 a) keep tension on a bronze pivot
 b) relieve pressure produced by adjusting nut
 c) avoid excessive pressure on valve when boiler is cold
2587. The recommended way to start a centrifugal pump with no head is:
 a) suction and discharge valves closed
 b) suction open and discharge closed
 c) suction closed and discharge open
2588. Which of the following is not a type of steam trap?
 a) bucket trap
 b) float and thermostatic trap
 c) P-trap
 d) impulse trap
2589. Which of the following is a type of steam heating system?
 a) direct expansion
 b) vacuum
 c) reverse return
 d) injection
2590. In the discussion of oil heating using a fuel pump, the number of pounds per square inch that pump pressure must drop to close nozzle valve is known as:
 a) delivery
 b) valve differential
 c) head of oil
 d) lift
2591. If a fuel system being checked is one in which the tank is buried below the burner level and the oil passes through a line filter, the _____ must show a reading; if it does not, an air leak is present.
 a) pressure gauge
 b) fuel gauge
 c) vacuum gauge
 d) sight glass
2592. What type of boiler has water in the tubes being heated by the hot gases outside the tubes?
 a) water tube boiler
 b) fire tube boiler
 c) radiant boiler
 d) atmospheric boiler
2593. Which one of these methods of heat transfer is not employed by the various types of heat transfer units?
 a) radiation

b) convection
c) gravitation
d) conduction

2594. Which of the following is considered a heating accessory?
a) boiler
b) piping
c) coils
d) pumps

2595. Which of the following is not a MAIN part of a heating system?
a) boiler
b) piping
c) heat transfer surfaces
d) water

2596. Which one of the following is a reason for a steam boiler getting too much water?
a) leaking hot water coil in boiler
b) faulty swing check in return header
c) trap plugged up
d) return pump not operating

2597. What is the weight of one cubic foot of water at 50 feet?
a) 40.52 lbs
b) 62.41 lbs
c) 60.33 lbs
d) 70.01 lbs

2598. When the pitch of the thread on a pipe is ¼ ", how many turns are required to thread 2 ½' of the pipe?
a) 8
b) 10
c) 12
d) 13

2599. What is the purpose of an expansion joint?
a) to minimize vibration
b) to allow for normal movement of the pipe due to expansion & contraction
c) to compensate for movements of pumps or compressors
d) to make removal of pipe easier

2600. What procedure should be followed when a beam interferes with the installation of a pipe line and the piping cannot be dropped below the beam and must be looped over the beam:
a) provide for an anchor to prevent vibration
b) provide for venting air from high point of pipe
c) use oversized fittings and pipe to prevent restriction
d) make sure that an expansion joint is used

2601. Which type of valve would be used in a horizontal line requiring complete drainage?
   a) globe
   b) check
   c) gate
   d) relief
2602. Which of the following valves should not be used for throttling purposes?
   a) plug
   b) needle
   c) gate
   d) globe
2603. All piping systems shall be capable of withstanding a dyfrostatic test pressure of how many times the designed pressure?
   a) 3
   b) 1 ½
   c) 2
   d) 2 ½
2604. Which of the following is not generally used to take care of thermal expansion in pipe lines?
   a) loop in the line
   b) packed slip-joints
   c) bellows-type joints
   d) change in size of line
2605. What type of metal would not be used on a safety valve seat?
   a) brass-copper and zinc
   b) carbon steel
   c) monel-nickel and copper
2606. Air in the oil line to burner may cause which of the following?
   a) pulsation
   b) excessive fuel consumption
   c) blue flame
   d) dirty nozzle
2607. A pipe which conducts condensation from the supply side to the return side of a steam heating system is:
   a) drip
   b) riser
   c) runout
   d) Hartford connection
2608. A pop safety valve on a boiler is usually provided with a hand lever, its purpose being:
   a) to close the valve once it pops open
   b) to relieve air
   c) manual safety
   d) to test the valve

2609. Steam mains and returns should be pitched not less than one inch per _____ feet in the direction of the steam flow.
   a) 10 ft
   b) 20 ft
   c) 30 ft
   d) 40 ft
2610. Each steam boiler shall have at least how many water glasses?
   a) 0
   b) 1
   c) 2
   d) 3
2611. The steam pressure gauge must be connected to the:
   a) steam space of the boiler
   b) main steam line
   c) water space of the boiler
2612. What is required between the steam gauge and the boiler?
   a) a trap or syphon with a valve of the "T" or lever handle type
   b) a vacuum breaker check valve
   c) cross fittings to facilitate cleaning
2613. A conventional float-operated low water fuel cut-off will turn off _____ but will not mechanically feed water to a boiler.
   a) a thermostatic fuel burner
   b) an automatic fuel burner
   c) an automatic water feeder
   d) an electric water feeder
2614. A steam trap is an automatic valve used in steam systems. Its total function is which of the following?
   a) permits passage of condensate
   b) permits passage of air
   c) permits passage of condensate, air and non-condensable gases
   d) permits passage of steam and air
2615. In a hot water heating system using a closed expansion tank, vents are required at:
   a) expansion tank only
   b) discharge side of pump
   c) last coil in system
   d) all high points
2616. In an automatically-fired steam boiler, the lowest safe water line with reference to the water glass is:
   a) top of water glass
   b) midway in water glass
   c) no lower than lowest visible part of the water glass
   d) 3/4 up from bottom of water glass

2617. A warm air furnace is sized for a space requiring 112,000 Btu/hr. The temperature rise of supply air over the return air is 100 degrees. The nearest cfm required for this condition is:
   a) 1,020 cfm
   b) 870 cfm
   c) 1,100 cfm
   d) 1,210 cfm
2618. On starting up a boiler using #2 fuel oil, a flue gas analysis shows 7% CO_2. How would this CO_2 reading be rated?
   a) excellent
   b) good
   c) fair
   d) poor
2619. One important function of primary control on an oil-fired furnace is to:
   a) open oil solenoid valve
   b) stop operation of unit in case of fire
   c) stop burner operation if flame is not established or if flame is extinguished after being in operation
   d) control oil pressure at burner nozzle
2620. A noisy heating system may be caused by:
   a) piping pockets
   b) boiling water in an open system
   c) floor holes too small for risers
   d) all of the above
2621. Correct safety control requires an ASME relief valve on:
   a) all high-pressure steam systems
   b) all low-pressure steam systems
   c) all ASME-rated systems
   d) every hot-water and steam-heating boiler
2622. An ASME relief valve should be connected:
   a) to the top of the boiler
   b) to the supply main at the boiler
   c) ahead of the fill-valve
   d) on the return main downstream of the pump
2623. Low-voltage controls are usually designed to operate at not over 25 volts.
   a) true
   b) false
2624. When a line voltage thermostat is used to control a hot water circulating pump, a relay is unnecessary unless the motor load exceeds the stat capacity.
   a) true
   b) false
2625. In oil-fired heat systems, the first control to act is the primary.
   a) true
   b) false

2626. Electric heating systems are thermally operated sequence controls or thermally operated staging controls as time-delay to energize the heating elements.
   a) true
   b) false
2627. Split-phase motors are only available in fractional horsepower; they use a centrifugal system which disconnects the starting windings as soon as the motor reaches operating speed.
   a) true
   b) false
2628. A capacitor-start motor develops a lower starting torque than a permanent split capacitor motor.
   a) true
   b) false
2629. When using a double element time-delay fuse with a 20 amp motor, the maximum fuse size should be:
   a) 60 amps
   b) 20 amps
   c) 30 amps
   d) 25 amps
2630. On a motor, a running capacitor is used to:
   a) increase torque
   b) increase horsepower
   c) increase power factor
   d) increase amperage
2631. What is the purpose of a thermostat?
   a) to control the heat
   b) to control the cold
   c) to cycle the cooling systems
   d) to control the temperature
2632. What is the purpose of the holding circuit in a magnetic motor starter?
   a) to hold the main contacts open until the control circuit through the interlock is made
   b) to hold the main contacts closed until the control circuit through the interlock is broken
   c) to hold the main contacts open until the control circuit through the interlock is broken
   d) to hold the main contacts closed until the control circuit through the interlock is made
2633. What will the readings of wet and dry hygrometers be if exposed to air that is completely saturated?
   a) they will both read alike
   b) the wet bulb will be lower
   c) the dry bulb will be lower

d) 90% relative humidity
2634. What type of electric motor has a set of field coils and a rotating armature?
   a) an induction motor
   b) a capacitor split-phase motor
   c) a repulsion-start induction-run motor
   d) a dual-winding repulsion-induction motor
2635. When air is passed through a water spray, some of the moisture of the air is given up only if the temperature of the water is:
   a) below the dewpoint of the air
   b) below the temperature of the air
   c) above the dewpoint of the air
   d) above the temperature of the air
2636. What factors determine the size of an expansion tank?
   a) the amount of space the water in the system requires in its expanded state
   b) the amount of space the air in the system requires in its expanded state
   c) the amount of air in the system
   d) the operating pressure in the system
2637. What does an anemometer measure?
   a) feet
   b) feet per minute (fpm)
   c) cubic feet (ft^3)
   d) cubic feet per minute (cfm)
2638. The term "induced draft" could refer to a type of:
   a) control
   b) compressor
   c) cooling tower
   d) diagram
2639. Which of the following is a device for removing dust from the air by means of electric charges induced on the dust particles?
   a) electric precipitator
   b) electric washer
   c) electric magnet
   d) electric ejector
2640. Which of the following is an instrument for measuring pressure, essentially a U-shaped tube partially filled with a liquid (usually water, mercury or a light oil) and constructed so that the amount of displacement of the liquid indicates the pressure being exerted on the instrument?
   a) potentiometer
   b) volometer
   c) manometer
   d) anemometer
2641. If condensate backs up in the return lines due to lack of proper head between the dry return and boiler water level, the water line in a steam boiler:
   a) rises several inches

b) rises
c) does not vary
d) drops

2642. The purpose of a Hartford Loop is to:
   a) prevent water from backing out of a boiler
   b) remove air from the return line
   c) allow for pipe expansion
   d) provide a balance for the steam header

2643. On a steam system, the operation of an inverted bucket trap is based on the:
   a) rise and fall of a float
   b) combination of steam pressure and weight of the condensate
   c) steam pressure drop
   d) weight of the water in the trap

2644. Where oil fuel tanks are lower than the burner on fuel-burning equipment, it is recommended that they have what kind of piping system?
   a) plastic pipe system
   b) 3-pipe system
   c) 2-pipe system
   d) 1-pipe system

2645. How many pounds of water can be heated 6° by 72 Btu?
   a) .833
   b) 12
   c) 432
   d) 72

2646. On a pneumatic control system, the main line pressure should be at how many psig?
   a) 25 psig
   b) 5 psig
   c) 15 psig
   d) 60 psig

2647. A pneumatic control valve which requires air pressure on the bellows or diaphragm to close the valve is a:
   a) diverting valve
   b) direct-acting valve
   c) reverse-acting valve
   d) normally closed valve

2648. A reverse-acting pneumatic humidistat increases air pressure to a controlled device when the humidity:
   a) increases
   b) decreases
   c) remains static
   d) mixes

2649. The bimetal strip in a thermostat is:
   a) the anticipator strip

b) the differential strip
c) the contactor strip
d) none of these

2650. On a pneumatic control system, the thermostat is connected to the valve of a damper motor by a small diameter tubing which is usually:
a) 1/4" OD
b) 3/8" OD
c) 3/8" ID
d) 1/4" ID

2651. For a heating application using a normally open valve, the controller is:
a) a submaster control
b) an indirect-acting run thermostat
c) a direct-acting run thermostat
d) a manual switch control

2652. The term applied to a device used to accumulate the effect of two or more thermostats to operate a single device is:
a) cumulator
b) accumulator
c) compensator
d) coordinator

2653. On a single-inlet blower fan, the driving side is on which side in relation to the inlet of the fan?
a) neither side
b) either side
c) opposite side
d) same side

2654. One of the three fundamental blower laws states that the power varies at the _____ of the speed.
a) rate
b) cube
c) square
d) circumference

2655. What is the size of grille or register to pass 800 cfm at 600 ft velocity (assume free area at 80%)?
a) 133 in^2
b) 197 in^2
c) 239 in^2
d) 314 in^2

2656. The unit of pressure commonly used in air ducts is measured in:
a) inches of water
b) inches of air
c) velocity
d) inches of mercury

2657. Which of the following electrical wires will carry the most current?
   a) 14
   b) 10
   c) 12
   d) 24
2658. An electrical network which has 120 volts to a neutral from all legs is called:
   a) Delta
   b) Star
   c) Polyphase
   d) 2-phase
2659. A squirrel cage fan is another name for:
   a) a backward inclined fan
   b) a forward inclined fan
   c) a radial flow fan
   d) an exhauster fan
2660. What is common to all filters?
   a) they must all be coated
   b) they are all constructed of fiberglass
   c) they are all 100% fireproof
   d) they all have a pressure drop
2661. What is the range of a thermostat?
   a) the open and close settings
   b) the temperature difference
   c) the capacity of the thermostat
   d) the variety of the models available
2662. By increasing a vent stack from 4" to 8", capacity will increase:
   a) 2 ½ times
   b) 4 times
   c) 2 times
   d) 3 times
2663. Which of the following would be a desirable velocity for a low-velocity system:
   a) 500 fpm
   b) 1,000 fpm
   c) 1,500 fpm
   d) 2,000 fpm
2664. Which would be a desirable friction loss per 100 ft of duct?
   a) .08
   b) .12
   c) .02
   d) .05
2665. Duct design criteria is not based on which of the following?
   a) friction loss
   b) velocity

c) Btu capacity
d) static pressure

2666. The weight of 26 ga galvanized sheet per square foot is:
a) .906 lbs
b) .708 lbs
c) 1.010 lbs
d) .560 lbs

2667. Uninsulated breeching for solid fuel burning furnace may not be within_____ inches of combustible material.
a) 18"
b) 10"
c) 24"
d) 6"

2668. If a gas-fired furnace has a 6" flue stack, what other exhaust may be connected to this stack?
a) toilet vent
b) kitchen range hood
c) laundry vent
d) no other

2669. In connecting duct work to a unit with two fan outlets, a common practice is to use a:
a) Pittsburgh lock
b) mixing box
c) pair of pants
d) splitter damper

2670. "Triangulation" is a term used in:
a) duct layout
b) duct insulation
c) duct sizing
d) none of the above

2671. A standing seam used as a cross seam for a large duct is:
a) used when a flat surface is required
b) used to eliminate the need for angle reinforcement
c) used for light gage duct
d) simple to make

2672. Describe a duct transition.
a) fitting for changing duct size or shape
b) fitting for splitter damper
c) fitting for branch take-off
d) fitting for fire damper

2673. A true grooved seam allowance is:
a) 1½ times the width of seam
b) 2 times the width of seam

c) 3 times the width of seam
   d) 1½ times the width of seam
2674. If a rectangular duct is 32" x 18", the equivalent in a round duct would be:
   a) 20"
   b) 26"
   c) 30"
   d) 22"
2675. In taking off a duct to determine weight, elbows are always measured:
   a) on center line
   b) twice
   c) on the radius
   d) on the perimeter
2676. Metal smoke stacks within 25 ft of any building must extend not less than:
   a) 8 ft
   b) 4 ft
   c) 2 ft
   d) 6 ft
2677. Flues for oil-fired warm air furnaces shall be:
   a) Class A
   b) Class D
   c) Class E
   d) Class W
2678. Mechanical ventilation for paint spray booths must have a:
   a) explosion proof motor
   b) two-speed motor
   c) back draft damper
   d) fusestat
2679. Which instrument shall be used to check motor performance on a mechanical ventilation system?
   a) pressure gauge
   b) ampmeter
   c) pitot tube
   d) velometer
2680. To check the rpm of a belt-driven fan, what instrument would be used?
   a) ampmeter
   b) tachometer
   c) velometer
   d) manometer
2681. To check air volume at a side wall supply grille, what instrument is used?
   a) tachometer
   b) velometer
   c) ampmeter
   d) mercury gauge

2682. Galvanized sheet metal is made by applying a protective coating of:
   a) lead
   b) zinc
   c) tin
   d) nickel

2683. To anneal copper:
   a) heat and cool slowly
   b) heat and quench
   c) apply electricity
   d) none of the above

2684. The following tool is used to cut out the cheeks of a 20 ga elbow:
   a) groover
   b) snips
   c) duck bills
   d) dividers

2685. When setting a seam by hand on a round pipe, use a:
   a) duck bill
   b) rivet set
   c) seams
   d) punch

2686. The Pittsburgh lock is used on:
   a) square ducts
   b) round pipe
   c) galvanized metal only
   d) expanded metal

2687. Duct work is bent up on a:
   a) ductulator
   b) slitter
   c) lock former
   d) brake

2688. Drive cleats can be made on a:
   a) seamer
   b) bar folder
   c) anvil or rail
   d) none of the above

2689. When drawing rivets, one would use:
   a) a rivet set
   b) a Whitney punch
   c) an awl
   d) a divider

2690. A duct system is considered high velocity when air travels in excess of:
   a) 12 mph
   b) 15,000 cfm

c) 800 gpm
d) 2500 fpm

2691. How many degrees are there in a semi-circle?
   a) 45°
   b) 90°
   c) 180°
   d) 360°

2692. In low-velocity air distribution systems, the flow of air to the branch take-offs is regulated by which of the following?
   a) amprobe
   b) volume controller
   c) splitter damper
   d) draft indicator

2693. Clean-out openings shall be provided in every metal smoke stack at:
   a) the base
   b) the furnace or heat producing apparatus
   c) specified intervals on the duct
   d) 6 ft from the floor

2694. A pipe fitting shaped like an ell, but with one female end and one male end, is called a:
   a) male union L
   b) female union L
   c) street L
   d) female-to-male L

2695. Threaded and tapped fittings that are screwed into the ends of other fittings or valves to reduce the size of the end openings are known as:
   a) reducers
   b) nipples
   c) bushings
   d) increasers

2696. What kind of threads does a street ell have?
   a) external threads only
   b) internal threads only
   c) both external and internal threads
   d) NF and NP threads

2697. For the ordinary installation, what is considered a good rule-of-thumb practice in spacing at intervals pipe hangers or supports for pipe?
   a) 5 ft
   b) 20 ft
   c) 10 ft
   d) 30 ft

2698. Which pipe has the thicker wall?
   a) Schedule 20
   b) Schedule 40

c) Schedule 80
d) Schedule 30

2699. A method of joining metal using fusible alloys having a melting point under 700° is:
a) acetelyne welding
b) brazing
c) soldering
d) arc welding

2700. A straight line drawn from the center to the extreme edge of a circle is known as the:
a) circumference of a circle
b) area of a circle
c) diameter of a circle
d) radius of a circle

2701. Zero on the Celsius thermometer is equivalent to what reading on the Fahrenheit scale?
a) 32°F
b) 100°F
c) 212°F
d) 0°F

2702. Using the conversion formula, -40°F would be equal to how many degrees on the Celsius scale?
a) +60°C
b) +10°C
c) -40°C
d) -60°C

2703. A good example of change in state is:
a) heating water from 32°F to 212°F
b) changing water from 32° ice to 32° liquid
c) heating steam at 212°F above its boiling point at atmospheric pressure
d) changing ice from 0°F to ice at 32°F

2704. The type of heat added in changing water from 32°F to water at 212°F is:
a) critical heat
b) latent heat
c) sensible heat
d) heat of fusion

2705. The amount of heat required to raise 10 lbs of water from 40°F to 100°F is:
a) 40 Btu
b) 400 Btu
c) 60 Btu
d) 600 Btu

2706. A good example of sublimation is the evaporation of:
a) water
b) dry ice

c) milk
  d) steam
2707. Water has a specific heat of 1.0 Btu/lb/°F. The amount of heat required to raise the temperature of 5 lbs of water from 10°F to 20°F is:
  a) 10 Btu
  b) 4.8 Btu
  c) 50 Btu
  d) 100 Btu
2708. The specific heat of water vapor is:
  a) 0.50
  b) 1.00
  c) 1.15
  d) 2.00
2709. Indoor velocities are usually measured in units of:
  a) miles per minute
  b) feet per minute
  c) meters per hour
  d) centimeters per second
2710. When heat flows upward from a register, heat is transferred by:
  a) vaporization
  b) convection
  c) conduction
  d) sublimation
2711. Comfort is:
  a) the feeling of contentment with the environment
  b) an atmosphere of 86 db, 40% rh
  c) air conditioning, like an ocean breeze
  d) warm, moist, clean air
2712. The body temperature is normally maintained at:
  a) 97° to 100°F
  b) 35°C
  c) 97° to 100°C
  d) 40° to 45°C
2713. To be comfortable, one must:
  a) wear warm clothing
  b) lose heat at the proper rate to the surrounding air
  c) be fairly active
  d) use only forced warm air heat
2714. The body transfers heat:
  a) by air that blows past it
  b) mainly by conduction
  c) by the normal body functions
  d) by radiation, convection and evaporation

2715. The amount of heat required to evaporate 1 lb of water at 212°F is approximately:
   a) 970 Btu
   b) 144 Btu
   c) 300 kcal
   d) 500 kcal
2716. Thermal comfort conditions include:
   a) freedom from noise
   b) freedom from disagreeable odors
   c) clean air
   d) temperature of air
2717. The ASHRAE Comfort Standard recommends which of the following temperature and relative humidity conditions:
   a) 75°F, 75% rh
   b) 76°F, 40% rh
   c) 80°F, 40% rh
   d) 27°C, 50% rh
2718. A psychrometer is:
   a) a chart measuring relative humidity
   b) a metric gauge for wire
   c) a device with 2 thermometers for measuring wet & dry bulb temperatures
   d) the name of a heating unit
2719. A psychrometric chart is:
   a) a device for measuring humidity
   b) a graph plotting air motion and temperature
   c) a chart where the various properties of air are plotted
   d) a chart showing how to assemble a heating unit
2720. An electronic filter will remove dust particles in which size range:
   a) 0.1 to 10 um
   b) 10 to 100 um
   c) .001 to 0.001 um
   d) 100 um or more
2721. Combustion is what type of process?
   a) mechanical
   b) chemical
   c) electrical
   d) pneumatic
2722. What are the three conditions necessary for combustion?
   a) fuel, air and moisture
   b) carbon, air and pressure
   c) fuel, heat and oxygen
   d) carbon, hydrogen and oxygen

2723. Complete combustion is obtained when all the carbon and hydrogen in the fuel are:
   a) oxidized
   b) heated
   c) combined
   d) dissipated
2724. Incomplete combustion may be caused by:
   a) a fire that is too hot
   b) too much draft
   c) insufficient fuel supply
   d) insufficient air supply
2725. What is the percent of oxygen in the air?
   a) 79%
   b) 21%
   c) 69%
   d) 31%
2726. Expressed in percentage, efficiency equals the useful heat divided by:
   a) the heat loss of the building
   b) the capacity of the furnace
   c) the heating value of the fuel
   d) the temperature of the flue gases
2727. The amount of excess air required for complete combustion is within which percentage range?
   a) 5 to 30%
   b) 25 to 60%
   c) 40 to 75%
   d) 5 to 50%
2728. The percent of CO_2 in the flue gas is not affected by the amount of excess air used by the combustion equipment.
   a) true
   b) false
2729. Natural gas is chiefly composed of:
   a) methane
   b) propane
   c) butane
   d) ethylene
2730. Is preheating of #6 oil usually necessary?
   a) true
   b) false
2731. Which of the following items is not included in a gravity-type furnace?
   a) heat exchanger
   b) cabinet
   c) air filter
   d) humidifier

2732. The location of the fan on a counterflow furnace is:
 a) above the heat exchange
 b) below the heat exchanger
 c) at the side of the heat exchanger
 d) in a separate unit
2733. The material used to construct the heat exchanger is:
 a) galvanized iron
 b) copper
 c) aluminum
 d) cold-rolled, low-carbon steel
2734. An individual-section type of heat exchanger is used on which kind of furnace?
 a) gas
 b) oil
 c) coal
 d) electric
2735. The type of fuel that requires the greatest chimney draft is:
 a) gas
 b) oil
 c) coal
 d) electric
2736. The fan blades on forced-air furnaces are:
 a) backward-curved
 b) forward-curved
 c) flat
 d) propeller-shaped
2737. If a motor has a 3" pulley, the fan a 6" pulley, and the motor runs at 1200 rpm, the speed of the fan is:
 a)   600 rpm
 b) 1,200 rpm
 c) 1,800 rpm
 d) 2,400 rpm
2738. If the fan is mounted on the shaft of the motor, the type of drive is called:
 a) unified
 b) V-belt
 c) indirect
 d) direct
2739. A downflow furnace is also called a:
 a) high-boy
 b) low-boy
 c) horizontal
 d) counterflow

2740. The colder the outside temperature, the greater or lesser the need for humidification?
   a) greater
   b) lesser
2741. The function of the gas burner assembly is to:
   a) produce proper fire at base of heat exchanger
   b) heat the air that goes to the room
   c) mix gas and air
   d) provide safe lighting of the burner
2742. A hand shut-off valve is used to turn gas on and off to the:
   a) burner
   b) pilot
   c) burner and pilot
2743. The proper gas manifold pressure for a residential gas unit is:
   a) 2½" wc
   b) 3½" wc
   c) 7" wc
   d) 11" wc
2744. A thermocouple produces how many millivolts of electric power?
   a) 25 to 30
   b) 35 to 50
   c) 500 to 600
   d) 750 to 800
2745. CGV stands for:
   a) continuous gas valve
   b) correct gas volume
   c) cast gas vent
   d) combination gas valve
2746. The maximum safety drop-out time for units with thermocouples is:
   a) 1 minute
   b) 2 minutes
   c) 3 minutes
   d) 4 minutes
2747. The type of pilot where gas is mixed with air before it enters the burner is:
   a) primary aerated
   b) non-primary aerated
2748. What type of fuel requires 100% shutoff at the gas valve?
   a) natural gas
   b) LP gas
2749. Air that is mixed with gas before burning is called:
   a) outside air
   b) total combustion air
   c) secondary air
   d) primary air

2750. A draft diverter:
- a) creates a draft of 0.02" of water column over fire
- b) connects directly to the gas burner
- c) is opened at the bottom
- d) is really not necessary on a modern furnace

2751. Oil is vaporized by:
- a) pressurization
- b) heat
- c) centrifugal force
- d) atomization

2752. The most popular oil burner is the:
- a) pot type
- b) rotary type
- c) low-pressure type
- d) high-pressure type

2753. A high-pressure burner mixes oil and air:
- a) before entering the nozzle
- b) after oil passes through the nozzle

2754. The electrodes on a gun-type burner are located:
- a) within the air/oil spray
- b) outside the air/oil spray

2755. The type of pump used with an outside tank located below the oil burner is called:
- a) two-stage
- b) single-stage

2756. The type of motor used on an oil burner is a:
- a) capacitor start
- b) split-phase
- c) capacitor run
- d) shaded pole

2757. The purpose of the nozzle on a high-pressure burner is to:
- a) blow air
- b) mix the oil and the air
- c) atomize the oil
- d) burn the oil

2758. The shape of the oil/air spray must:
- a) correspond to the type of combustion chamber
- b) fit the burner supplied
- c) make a good-appearing fire
- d) fit the heating load

2759. The oil capacity (in gallons) of a single inside tank is usually:
- a) 275
- b) 550
- c) 1,000

d) 1,500

2760. What size copper tubing is usually used for inside tank oil lines?
   a) 1/4" OD
   b) 3/8" OD
   c) 1/2" OD
   d) 5/8" OD

2761. The products of combustion are:
   a) C SO_2 H_2O N_2 H
   b) CO_2 NaCe $AgNO_3$
   c) CO_2 CO O N H_2O
   d) AgCl KCl C_6H_5

2762. In practice, a gas burner should be adjusted to produce what percent of CO_2 in the flue gas?
   a) 10 to 12%
   b) 8.75 to 9.5%
   c) 4 to 7.5%
   d) 20 to 25.5%

2763. An instrument to check CO_2 is a:
   a) flue gas analyzer
   b) true spot tester
   c) micrometer
   d) psychrometer

2764. A burner should be run for a minimum of how many minutes before combustion tests are made?
   a) 1 to 2
   b) 5 or 10
   c) 15 to 20
   d) 30 to 45

2765. What is the maximum number of holes that should be made in the flue for combustion tests?
   a) one
   b) two
   c) three
   d) four

2766. Draft at the burner (operating) should be:
   a) negative
   b) positive
   c) neutral

2767. The proper reading for the smoke test for a new oil furnace should be:
   a) number 1 or 2
   b) number 3 or 4
   c) number 5 or 6
   d) number 7 or 8

2768. In making the CO_2 test on an oil furnace, the bulb should be collapsed how many times?
   a) six
   b) eight
   c) twelve
   d) eighteen
2769. Combustion efficiency should be:
   a) 50% or better
   b) 60% or better
   c) 75% or better
   d) 85% or better
2770. If the combustion efficiency is 80%, what should the stack loss be?
   a) 80%
   b) 20%
   c) 70%
   d) 30%
2771. Which of the following is not a load device?
   a) fan
   b) humidifier
   c) heater
   d) fan control
2772. Which basic element of the furnace control system includes the sensor?
   a) power supply
   b) gas valve
   c) limit devices
   d) fan motor
2773. What is the smallest variation in degrees Fahrenheit that people can sense?
   a) ½
   b) 1½
   c) 2½
   d) 4
2774. A low-voltage control system operates on how many volts?
   a) 24
   b) 30
   c) 120
   d) 240
2775. On a bimetallic thermostat, the metal used in addition to copper is:
   a) zinc
   c) invar
   b) chromium
   d) iron
2776. What is the largest number of switches commonly found in a single mercury tube element?
   a) one

b) two
c) three
d) four

2777. A self-generating system usually uses its own control circuit of how many volts or millivolts?
   a) 30 mv
   b) 750 mv
   c) 24 V
   d) 120 V

2778. If a thermostat cuts in at 70°F and cuts out at 72°F, the differential is:
   a) zero
   b) 2°F
   c) 70°F
   d) 72°F

2779. A heating system should cycle how many times per hour?
   a) 1 to 3
   b) 4 to 7
   c) 8 to 10
   d) 11 to 15

2780. The color code terminals of a thermostat are:
   a) Q, H, C, F
   b) R, G, B, V
   c) T, A, E, H
   d) R, W, Y, G

2781. A timed fan start delays the operation of the fan approximately how many seconds after the heater is turned on:
   a) 30
   b) 60
   c) 90
   d) 120

2782. A limit control is usually set to cut out the source of heat at how many Fahrenheit degrees?
   a) 180°
   b) 200°
   c) 220°
   d) 240°

2783. What size of branch circuit can usually be recommended by code for heating furnace?
   a) 10 A
   b) 15 A
   c) 20 A
   d) 25 A

2784. How many circuits are required to operate a two-speed fan?
   a) one
   b) two
   c) three
   d) four
2785. On a call for heating, which two terminals on the thermostat close (make)?
   a) R and W
   b) R and Y
   c) R and B
   d) G and W
2786. With a cut-out temperature of 200°F for the limit control, what would the normal cut-in be?
   a) 150°F
   b) 175°F
   c) 200°F
   d) 225°F
2787. What is the usual cut-out temperature for the secondary limit control?
   a) 145°F
   b) 160°F
   c) 175°F
   d) 200°F
2788. When the thermostat calls for cooling, what two terminals on the thermostat close (make)?
   a) R and W
   b) R and Y
   c) R and G
   d) G and W
2789. What control device activates a timed fan start?
   a) gas valve
   b) limit control
   c) thermostat
   d) fan control
2790. What accessory can be added that requires a second power supply?
   a) humidifier
   b) cooling
   c) electrostatic air cleaner
   d) stoker
2791. What are the new thermostat terminals required when a two-stage gas valve arrangement is used?
   a) R1 and R2
   b) W1 and W2
   c) b1 and b2
   d) Y1 and Y2

2792. In testing a fan relay, what voltage is usually applied to the relay coil?
   a) 24 V
   b) 120 V
   c) 240 V
   d) 480 V
2793. On the stack relay type of oil primary control, are the cold contacts normally open or normally closed?
   a) normally open
   b) normally closed
2794. What is the light-sensitive material used on a cad cell?
   a) cadmium nitrate
   b) copper sulfate
   c) sodium chloride
   d) cadmium sulfate
2795. What is the resistance of a cad cell in the absence of light?
   a) 5,000
   b) 50,000
   c) 100,000
   d) 150,000
2796. To which terminals on the primary control is the cad cell usually connected?
   a) T and S
   b) S and S
   c) R and W
   d) W and R
2797. How many low-voltage terminals on a cad cell primary control?
   a) two
   b) four
   c) six
   d) eight
2798. On a cad cell primary control, the "hot" side of the power supply is wired to what connection?
   a) black
   b) orange
   c) blue
   d) white
2799. On the oil burner stack relay, the power supply is wired to which terminals?
   a) 0 and 1
   b) 1 and 2
   c) 2 and 3
   d) 3 and 4
2800. In checking a switch with a voltmeter, a reading of zero indicates:
   a) a closed switch
   b) an open switch

c) a short
d) a ground

2801. The specific gravity of natural gas is:
   a) 0.65
   b) 1.00
   c) 1.50

2802. If the pilot gas goes out on gas burners having an hourly input of less than 400,000 Btu, the main gas shall shut off within:
   a) 10 seconds
   b) 50 seconds
   c) 3 minutes

2803. Combustion Efficiency is determined from:
   a) stack temperature and percentage of CO
   b) stack temperature and inches of stack draft
   c) stack temperature and smoke spot

2804. The heat value of natural gas is:
   a) 500 Btu/ft^3
   b) 1,000 Btu/ft^3
   c) 2,500 Btu/ft^3

2805. Minimum size of a hand hole plate is:
   a) 3 x 4
   b) 2 3/4 x 3 1/2
   c) 2 x 4

2806. The limit setting on a standard efficiency forced warm air furnace would be:
   a) 140°F
   b) 200°F
   c) 300°F

2807. When the heating system will not work with the thermostat jumpered, what problem is indicated?
   a) a bad thermostat
   b) something other than the thermostat
   c) a definite electrical problem

2808. A good thermocouple will read open circuit:
   a) 4-7 mv
   b) 12-17 mv
   c) 17-30 mv

2809. A good pilotstat power unit will operate in the range of:
   a) 4-7 mv
   b) 12-17 mv
   c) 17-30 mv

2810. A good thermocouple will read in a loaded circuit:
   a) 4-7 mv
   b) 12-17 mv
   c) 17-30 mv

2811. Continuous blower operation can be used for:
   a) cooling only
   b) both heating and cooling
   c) heating only
2812. A DPDT blower relay (fan center) will protect the system from:
   a) inadequate speed control
   b) backfeed through multi-speed motor windings
   c) overloading the multi-speed motor
2813. The gas pilot flame should envelop the thermocouple around the:
   a) top 1 inch
   b) top 3/4 inch
   c) top 1/2 inch
2814. The orifice opening in an LP gas-fired furnace is _____ for the same Btu rating.
   a) smaller than for natural gas
   b) larger than for natural gas
   c) the same size as for natural gas
2815. Normal natural gas pressure from the meter is:
   a) 6 to 7" wc
   b) 3.5" wc
   c) 11" wc
2816. Allowable pressure loss due to pipe friction is:
   a) 0.3" wc
   b) 0.5" wc
   c) 0.65" wc
2817. When checking for gas leaks, use:
   a) soapy water
   b) safety matches
   c) electronic leak detector
2818. Concealed joints in gas piping must be:
   a) painted yellow where they come out of the wall
   b) inspected before covering
   c) covered with anti-leak compound
2819. If units at the highest points of a hot-water heating system are found to be not hot, these units are most likely to be:
   a) undersized
   b) air-bound
   c) on a vacuum
2820. When operating a fuel oil furnace with a one-pipe system on the fuel pump:
   a) the bypass plug should be installed
   b) the single pipe should be teed into both suction parts
   c) the bypass plug should be removed
2821. A gas drip leg must:
   a) be installed in humid or wet areas

b) have the capacity to catch 150% of the drippings
c) not be smaller in diameter than the pipe it connects to

2822. Gas piping cannot be run through:
   a) unvented basements
   b) other apartments
   c) return ducts and clothes chutes

2823. A thermostat that cycles too often or not enough indicates:
   a) thermostat not mounted level
   b) defective or burnt heat anticipator
   c) improper heat anticipator setting

2824. Recuperative furnaces have an efficiency of:
   a) 70%
   b) 80%
   c) 90% +

2825. Cycle pilot furnaces have an efficiency of:
   a) 70% +
   b) 80% +
   c) 90% +

2826. Condensing furnaces have an efficiency of:
   a) 70% +
   b) 80% +
   c) 90% +

2827. The main difference between recuperative and condensing furnaces is in:
   a) the secondary heat exchanger
   b) the control system
   c) the vent system

2828. Furnace flue gas condensate is:
   a) basic
   b) neutral
   c) acidic

2829. The maximum flue gas temperature for furnaces utilizing PVC pipe is:
   a) 100°F
   b) 140°F
   c) 180°F

2830. On a call for heat, the first step on a mechanical vent furnace is:
   a) burner operation
   b) blower operation
   c) combustion chamber purge cycle

2831. Most condensing furnaces may not be connected to the existing chimney unless:
   a) stainless steel connector pipe is used
   b) PVC connector pipe is used
   c) the furnace manufacturer permits this application

2832. Redundant gas valves utilizing flame rectification to prove a pilot flame must be grounded:
   a) to earth ground
   b) to the neutral ground bar in the fuse box
   c) to any ground available

2833. To prove a pilot flame, most White Rodgers redundant gas valves use a:
   a) bimetal sensor
   b) mercury vapor sensor
   c) flame rod (rectification) sensor

2834. To prove a pilot flame, most Bryant & Carrier redundant gas valves use a:
   a) bimetal sensor
   b) mercury vapor sensor
   c) flame rod (rectification) sensor

2835. To prove a pilot flame, most Honeywell redundant gas valves use a:
   a) bimetal sensor
   b) mercury vapor sensor
   c) flame rod (rectification) sensor

2836. The redundant gas valve with the quickest response in sensing a pilot flame and energizing the main gas valve utilizes the:
   a) bimetal sensor
   b) mercury vapor sensor
   c) flame rod (rectification) sensor

2837. During a call for heat, the redundant gas valves with Pick and Hold coils continuously energize:
   a) the Pick coil
   b) the Hold coil
   c) neither coil

2838. The 100% lockout switch is opened when:
   a) the pilot flame is not established within a defined time limit
   b) the gas valve is not energized within a defined time limit
   c) the combustion vent fan is not proven

2839. The combustion vent fan proving switch operates between:
   a) -0.03 to -0.06" wc
   b) -0.3 to - 0.6" wc
   c) -3.0 to - 6.0" wc

2840. On redundant gas valves, the thermostat heat anticipator setting will be:
   a) higher than with standard gas valves
   b) lower than with standard gas valves
   c) the same as with standard gas valves

2841. Excessive sooting in the heat exchangers of a condensing furnace may be caused by:
   a) high gas pressure
   b) partial blocking by condensate
   c) an inoperative flue gas vent fan

2842. To check for dirty exterior fin tubing on a secondary heat exchanger, check for:
   a) high flue gas temperatures
   b) high temperature rise through the furnace
   c) low gas pressure drop
2843. To check for plugged interior tubing on a secondary heat exchanger, check:
   a) high flue gas temperatures
   b) high temperature rise through the furnace
   c) low gas pressure drop
2844. Two basic types of gas burner ignition systems are:
   a) thermal and mercury
   b) direct and indirect
   c) lockout and non-lockout
2845. Lennox Pulse furnaces have controlled explosions at the rate of:
   a) 50 to 70 cycles/second
   b) 50 to 70 cycles/minute
   c) 50 to 70 cycles/hour
2846. When reducing the input of a gas furnace to more closely represent the house heat loss, you should:
   a) reduce the orifice size by no more than 20%
   b) reduce the gas pressure by no more than 20%
   c) either of the above
2847. Natural gas burner orifices are:
   a) larger in diameter than propane gas burner orifices
   b) smaller in diameter than propane gas burner orifices
   c) the same diameter as propane gas burner orifices
2848. A smoke spot of 5 will produce soot that:
   a) is extremely light if there is any at all
   b) will require cleaning once per year
   c) will collect heavily and rapidly
2849. Excessive draft can:
   a) increase the stack temperature
   b) reduce the percentage of CO_2 in the flue gases
   c) all of the above
2850. The hole for the stack thermometer should be:
   a) ¼ inch in diameter
   b) 6 inches from and on the furnace side of the draft regulator
   c) all of the above
2851. Natural gas pressure at the outlet of the gas meter is usually:
   a) 3.5" wc
   b) 7.0" wc
   c) 11.0" wc
   d) 14.7" wc

2852. Natural gas pressure within the burner manifold is usually:
   a) 3.5" wc
   b) 7.0" wc
   c) 11.0" wc
   d) 14.7" wc
2853. The purpose of the "standing pilot" is to:
   a) heat the thermocouple
   b) ignite the main burner
   c) generate a millivoltage
   d) both a and c
2854. The output voltage of a thermocouple may be as high as:
   a) 30 mv
   b) 250 mv
   c) 500 mv
   d) 750 mv
2855. The pressure regulator on an LP system is set at:
   a) 3.5" wc
   b) 7.0" wc
   c) 11.0" wc
   d) 14.7" wc
2856. The output voltage of a thermopile (PowerPile) may be as high as:
   a) 30 mv
   b) 250 mv
   c) 500 mv
   d) 750 mv
2857. The pilot flame should envelop the end of the thermocouple:
   a) 1/8 to 3/8 inch
   b) 3/8 to 1/2 inch
   c) 1/2 to 3/4 inch
   d) 1/8 to 3/4 inch
2858. Primary air is mixed with the gas within the:
   a) manifold
   b) orifice
   c) venturi
   d) burner head
2859. On flanged joints, bolts should be tightened in which manner?
   a) crossover method
   b) rotation
   c) by hand
   d) welded
2860. The Pittsburg lock is used on:
   a) square ducts
   b) round pipe

c) galvanized metal only
d) expanded metal

2861. A yellowish flame at the burner is usually caused by:
   a) too little primary air
   b) too much primary air
   c) the secondary air ratio factor
   d) the primary air dilution source

2862. Manifold gas pressure that is too low could cause the flame to:
   a) flash back through the burner
   b) lift off the burner
   c) reduce burner efficiency
   d) dilute the secondary air ratio factor

2863. Excessive primary air could cause the flame to:
   a) flash back through the burner
   b) lift off the burner
   c) reduce burner efficiency
   d) dilute the secondary air ratio factor

2864. The air that is mixed with the gaseous fuel before ignition is called:
   a) mixed air
   b) primary air
   c) secondary air
   d) oxygen

2865. Combustion efficiency is established through:
   a) percent CO_2
   b) flue gas temperatures
   c) all of the above
   d) none of the above

2866. Average gas burning combustion efficiencies are:
   a) 40% to 60%
   b) 50% to 65%
   c) 77% to 80%
   d) 77% to 95%

2867. The function of the gas burner assembly is to:
   a) produce proper fire at the base of the heat exchanger
   b) heat the air that goes to the rooms
   c) mix gas and air
   d) provide safe lighting of the burner

2868. The type of gas valve operator with a delayed action feature is a:
   a) solenoid
   b) diaphragm
   c) bimetal
   d) bulb

2869. A "fast acting" type of gas valve is the:
   a) solenoid

b) diaphragm
c) bimetal
d) bulb

2870. A draft diverter:
a) creates a draft of 0.02 inches of water column over the fire
b) connects directly to the gas burner
c) is open at the bottom
d) is really not necessary on an atmospheric burner

2871. On a call for heating, which two terminals on the thermostat close?
a) R & W
b) R & Y
c) R & G
d) G & W

2872. On a call for cooling, which two terminals on the thermostat close?
a) R & W
b) R & Y
c) R & G
d) G & W

2873. An oil-fired unit using No. 2 fuel oil that burns 105,000 Btu would have a _____ gph nozzle.
a) .65
b) .75
c) .85
d) .90

2874. To which terminal on the primary control is the Cad-cell connected?
a) CR
b) RW
c) TT
d) FF

2875. What is the usual cut-out temperature for the secondary limit control on a horizontal furnace?
a) 145°F
b) 160°F
c) 175°F
d) 200°F

2876. Which type of oil burner mixes the oil and air before entering the nozzle?
a) low-pressure
b) rotary
c) high-pressure
d) pot type

2877. Water-gauge glasses must have a blow down valve with a ____ inch minimum diameter opening.
a) 1
b) 3/4

c) 1/2
d) 1/4
2878. What is the approximate resistance of a cad cell in the presence of light?
   a) 100 ohm
   b) 1,000 ohm
   c) 10,000 ohm
   d) 100,000 ohm
2879. On a cad cell primary control, the "hot" line is connected to the:
   a) orange lead
   b) black lead
   c) brown lead
   d) white lead
2880. On the stack primary control, line power is connected to terminals:
   a) 1 & 2
   b) 2 & 3
   c) 3 & 4
   d) T & T
2881. If the flame fails, the cad cell relay will lock out on safety in about:
   a) 30 sec
   b) 60 sec
   c) 90 sec
   d) 120 sec
2882. If the flame fails, the stack relay will lock out on safety in about:
   a) 30 sec
   b) 60 sec
   c) 90 sec
   d) 120 sec
2883. Oil primary relays are usually wired directly to the:
   a) line power
   b) fan switch
   c) limit switch
   d) transformer
2884. On an oil primary relay (the constant ignition type), the oil burner motor and the ignition transformer are wired in:
   a) series
   b) parallel
   c) combination series and parallel
   d) either series or parallel
2885. The secondary voltage of an oil burner ignition transformer is:
   a) 1,000 V
   b) 5,000 V
   c) 10,000 V
   d) 100,000 V

2886. The stack primary control should be installed in the flue pipe between:
   a) the furnace and the barometric damper
   b) the barometric damper and the chimney
   c) heat exchanger sections
   d) the burner housing and the flame
2887. The cad cell sensor is located in the:
   a) burner housing
   b) cad cell relay
   c) combustion chamber
   d) stack relay
2888. On a counterflow-type oil furnace, the auxiliary limit switch and the normal limit switch are wired in:
   a) series
   b) parallel
   c) combination series and parallel
   d) either series or parallel
2889. If oil in suspension with air is allowed to touch a cold surface, it will:
   a) heat that cold surface
   b) condense into a liquid
   c) produce a flash fire
   d) burn with a cracking sound
2890. Assuming perfect combustion, the percent of CO_2 found in flue gas is:
   a) 4 to 6%
   b) 6 to 10%
   c) 10 to 12%
   d) 15%
2891. Assuming normal good combustion, approximately what percentage of CO_2 should be found in the flue gas?
   a) 4 to 6%
   b) 6 to 10%
   c) 10 to 12%
   d) 15%
2892. When the flame is being well utilized, there should be:
   a) high stack temperature
   b) low stack temperature
   c) cold stack temperature
   d) diluted stack temperature
2893. Excessive draft will result in:
   a) high stack temperature
   b) low stack temperature
   c) cold stack temperature
   d) diluted stack temperature
2894. The purpose of the nozzle on a high pressure burner is to:
   a) blow air

b) mix the oil and the air
c) atomize the oil
d) burn the oil

2895. To provide better lift from the tank to the burner, you need a:
    a) single-stage pump
    b) two-stage pump
    c) single-pipe system
    d) two-pipe system

2896. To provide better removal of air in the oil supply, you should have a:
    a) single-stage pump
    b) two-stage pump
    c) single pipe system
    d) two pipe system

2897. The electrodes on a gun-type oil burner are located:
    a) within the air/oil spray
    b) outside of the air/oil spray
    c) on the ignition transformer
    d) in front of the nozzle

2898. A boiler which has a self-supporting water-cooled shell or furnace bottom is called a _____ boiler.
    a) wet bottom
    b) wet back
    c) water-cooled
    d) water tube

2899. On the bimetal stack relay, which terminal is for the burner motor?
    a) 4
    b) 2
    c) 1
    d) 3

2900. A time start fan control is usually used on:
    a) a counterflow furnace
    b) a downflow furnace
    c) a horizontal furnace
    d) any of the above

2901. High-pressure steam starts at:
    a) 30 lbs
    b) 125 lbs
    c) 3 to 5 lbs
    d) none of the above

2902. Fuel oil storage, attached to the oil-fired room heater by the manufacturer, is not to exceed:
    a) 20 gal
    b) 10 gal

c) 5 gal
d) 15 gal

2903. On a low-pressure "residential" forced-water boiler, the relief device would open up at:
a) 125 lbs
b) 15 lbs
c) 30 lbs
d) none of the above

2904. On a "residential" low-pressure steam boiler, the pressure reducing valve would be factory set at:
a) 30 lbs
b) 112 lbs
c) 12 lbs
d) none of the above

2905. On a low-pressure hot water boiler, maximum temperature is not to exceed:
a) 210°F
b) 160°F
c) 250°F
d) 212°F

2906. A permit is not required on gas burners with inputs of less than:
a) 50,000 Btu
b) 40,000 Btu
c) 30,000 Btu
d) 20,000 Btu

2907. Roof-mounted equipment not exceeding 400 lbs should be supported by at least how many wood joists?
a) 1
b) 2
c) 3
d) 4

2908. Roof-mounted equipment not exceeding 400 lbs should be supported by at least how many trusses?
a) 1
b) 2
c) 3
d) 4

2909. What is the maximum allowable percentage above the rated input Btu at which a gas furnace can be safely operated?
a) 20%
b) 10%
c) 0%
d) none of the above

2910. The clearance in front of an oil furnace for a typical municipalitiy is:
   a) 6"
   b) 24"
   c) 18"
   d) none of the above
2911. The ventilation and air for combustion from inside a building must be at least one square inch per:
   a) 1,000 Btu
   b) 2,000 Btu
   c) 3,000 Btu
   d) 4,000 Btu
2912. The ventilation and air for combustion from outdoors through horizontal ducts must be at least one square inch per:
   a) 1,000 Btu
   b) 2,000 Btu
   c) 3,000 Btu
   d) 4,000 Btu
2913. The ventilation and air for combustion from outdoors through vertical ducts must be at least one square inch per:
   a) 1,000 Btu
   b) 2,000 Btu
   c) 3,000 Btu
   d) 4,000 Btu
2914. The setting of the limit control on a mechanical warm air furnace shall not exceed:
   a) 100
   b) 200
   c) 300
   d) 212
2915. The setting of the limit control on a hot water supply heater or boiler shall not exceed:
   a) 125
   b) 212
   c) 210
   d) 250
2916. The setting of the limit control on a low-pressure hot water space-heating boiler shall not exceed:
   a) 125
   b) 212
   c) 210
   d) 250

2917. Minimum oil pipe size for domestic-type burners would be:
   a) 1/2"
   b) 3/8"
   c) 3/4"
   d) 1"
2918. Oil tank vent pipes are:
   a) 1 1/2" bip
   c) 1 3/4" bip
   b) 1 1/4" bip
   d) 1 7/8" bip
2919. Oil tank fill pipes are:
   a) 1 1/2" bip
   c) 1 3/4" bip
   b) 1 1/4" bip
   d) 1 7/8" bip
2920. The gauge thickness of an oil tank installed indoors would be:
   a) 10 ga
   b) 12 ga
   c) 14 ga
   d) 16 ga
2921. An oil tank vent pipe must terminate not less than _____ from building openings.
   a) 1 ft
   b) 1.5 ft
   c) 2 ft
   d) 3 ft
2922. The difference between the opening and closing pressure of a safety or relief valve is known as the:
   a) differential pressure
   b) pressure drop
   c) blow down
   d) none of the above
2923. On oil-fired units, with #2 oil and less than 5 gal, the burner will shut off within:
   a) 60 sec
   b) 10 sec
   c) 2 min
   d) 3 min
2924. Normal operating pressure for residential forced-water boilers would be:
   a) 30 lbs
   b) 12 to 15 lbs
   c) 3 to 5 lbs
   d) none of the above

2925. Normal operating pressure for residential steam boilers would be:
- a) 3 to 5 lbs
- b) 12 to 15 lbs
- c) 30 lbs
- d) none of the above

2926. The number of Btu for one cubic foot of natural gas would be:
- a) 2,500 Btu
- b) 1,000 Btu
- c) 84,000 Btu
- d) 100,000 Btu

2927. The number of Btu for one gallon of #2 oil would be:
- a) 100,000 Btu
- b) 137,000 Btu
- c) 84,000 Btu
- d) 2,500 Btu

2928. Return air grilles for perimeter heating systems may be located in the:
- a) floor and baseboard
- b) ceiling
- c) high in the sidewall
- d) all of the above

2929. How much air is needed to burn 1,000 Btu?
- a) 100 ft^3
- b) 50 ft^3
- c) 10 ft^3
- d) 1 ft^3

2930. The pressure from the meter to the appliance would be:
- a) 6 to 7" wc
- b) 3 to 4" wc
- c) 11" wc
- d) none of the above

2931. The manifold pressure of natural gas would be:
- a) 6 to 7" wc
- b) 11" wc
- c) 3 to 4" wc
- d) none of the above

2932. What is the permissible pressure drop between the meter and the appliance?
- a) 3.0" wc
- b) 0.05" wc
- c) 5.0" wc
- d) none of the above

2933. Underground tanks for storing flammable liquids must have a vent pipe draining to the tank; the top of such vent pipe shall not be closer than _____ feet to any building.
- a) 1 ft

b) 3 ft
c) 5 ft
d) 10 ft

2934. Duct systems conveying exhaust from restaurant hoods shall operate at a velocity no less than _____ fpm.
   a) 1,000 fpm
   b) 1,500 fpm
   c) 1,700 fpm
   d) 1,850 fpm

2935. Metal smoke stacks 12 to 16" in diameter shall be not less than _____ gauge.
   a) 24 ga
   b) 22 ga
   c) 12 ga
   d) 10 ga

2936. The flue pipe from a gas designed furnace shall be at least:
   a) 8"
   b) 6"
   c) 5"
   d) none of the above

2937. A high velocity round duct 23 inches in diameter with a longitudinal seam construction must be at least _____ gauge sheet steel.
   a) 14 ga
   b) 16 ga
   c) 18 ga
   d) 20 ga

2938. A low-water cutoff is a de-energizer set to trip when the:
   a) low water reaches 180°F
   b) low water rises
   c) boiler water drops
   d) none of the above

2939. A boiler operating at a pressure greater than 100 psi, the blow-off piping thickness shall not be less than:
   a) standard black iron pipe (bip)
   b) heavy copper pipe
   c) schedule 80 pipe
   d) schedule 40 pipe

2940. An ASME pressure relief valve on a boiler is usually provided with a hand lever. Its primary purpose is to:
   a) close the valve once it has opened
   b) test the valve and relief of air on initial fill
   c) change operating pressure
   d) provide manual safety

2941. A steam or hot water boiler must have a:
   a) drain valve

b) individually controlled make up valve
   c) both a and b
   d) none of the above
2942. Which valve is most suitable for throttling flow:
   a) gate
   b) check
   c) globe
   d) swing
2943. What is the normal wattage capacity of a 120V electric heating thermostat?
   a) 100 watts
   b) 1,000 watts
   c) 3,000 watts
   d) 5,000 watts
2944. What voltage is used on the limit control of a gun-type oil burner system?
   a) 24 V
   b) 115 V
   c) 30 mv
   d) 750 mv
2945. Which pipe has the thickest wall?
   a) schedule 20
   b) schedule 40
   c) schedule 60
   d) schedule 80
2946. Clean-out openings shall be provided in every metal smoke stack at:
   a) the furnace or heat-producing apparatus
   b) the base
   c) 6 ft from the floor
   d) none of the above
2947. Chimneys for oil-fired warm air furnaces shall be:
   a) class A or B
   b) class A or C
   c) class B or C
   d) class B
2948. One boiler horsepower is equal to:
   a) 144 Btu
   b) 970 Btu
   c) 33,475 Btu
   d) 134,475 Btu
2949. An oil tank of 500 gallons must be what gauge?
   a) 10 ga
   b) 12 ga
   c) 14 ga
   d) 16 ga

2950. What is the minimum allowable percentage of Btu at which a gas furnace can safely be fired?
   a) 80%
   b) 90%
   c) 100%
   d) none of the above
2951. The effective height of a chimney is:
   a) the vertical distance from point of entry to the chimney to the ceiling
   b) the vertical distance from point of entry to the chimney to the roof
   c) the vertical distance from point of entry to the chimney to roof peak
   d) the vertical distance from point of entry to the chimney to the top of the chimney
2952. A limiting device installed on a steam boiler is controlled by:
   a) a vent
   b) temperature
   c) pressure
   d) hydraulic
2953. A limiting device installed on a hot water boiler is controlled by:
   a) vent
   b) temperature
   c) pressure
   d) hydraulic
2954. A limiting device installed on a forced air furnace is controlled by:
   a) vent
   b) temperature
   c) pressure
   d) hydraulic
2955. Gas burners having an hourly input of 400,000 Btu will have a trial for ignition period not exceeding:
   a) 2 min
   b) 3 min
   c) 60 sec
   d) 10 sec
2956. Vent connectors and chimney connectors shall be at least:
   a) 26 ga
   b) 28 ga
   c) 30 ga
   d) 14 ga
2957. The flame failure timing period on furnaces whose firing rate is 5 gph is:
   a) 2 min
   b) 3 min
   c) 10 sec
   d) 60 sec

2958. Ducts shall be supported no further apart than:
- a) 5 ft
- b) 8 ft
- c) 10 ft
- d) 12 ft

2959. 1" gas pipe shall be supported no further apart than:
- a) 5 ft
- b) 8 ft
- c) 10 ft
- d) 12 ft

2960. The minimum clearance in front of a gas furnace is:
- a) 6"
- b) 12"
- c) 18"
- d) 24"

2961. The minimum clearance in front of an oil furnace is:
- a) 6"
- b) 12"
- c) 18"
- d) 24"

2962. The minimum clearance of the top of an oil furnace is:
- a) 6"
- b) 12"
- c) 18"
- d) 24"

2963. The minimum clearance of the side of a gas furnace is:
- a) 6"
- b) 12"
- c) 18"
- d) 24"

2964. When equipment is installed in a place such as a roof, supplying safe access and approach to this equipment is the responsibility of:
- a) the building owner
- b) the heating contractor
- c) the building contractor
- d) none of the above

2965. Openings in building construction to equipment shall be at least:
- a) 24" x 36"
- b) 18" x 36"
- c) 24" x 24"
- d) 36" x 36"

2966. Natural gas pressure at the street is usually:
- a) 15 to 30" wc
- b) 15 to 60 lbs

c) 5 to 30 lbs
d) 15 to 60" wc

2967. Propane manifold pressure is:
    a) 3 to 5" wc
    b) 6 to 7" wc
    c) 5 to 7" wc
    d) none of the above

2968. Propane tank pressure is:
    a) 15 to 30 lbs
    b) 50 to 100 lbs
    c) 100 to 200 lbs
    d) 200 to 300 lbs

2969. The three types of pilot safety devices are:
    a) current, hydraulic, bimetal
    b) current, motorized, diaphragm
    c) current, potential, bimetal
    d) voltage, hydraulic, bimetal

2970. The three types of gas valves are:
    a) current, motorized, potential
    b) current, motorized, diaphragm
    c) current, hydraulic, bimetal
    d) motorized, current, bimetal

2971. Chimney heights for #2 oil or gas-fired equipment of 500,000 Btu or less are ____ ft above flat roofs, and at least ____ ft above ridges, peaks, etc. within ____ ft.
    a) 3, 2, 10
    b) 1, 2, 3
    c) 2, 3, 3
    d) 2, 3, 10

2972. The maximum length of chimney connecter is:
    a) 3/4 of the distance from the chimney entrance to the chimney top
    b) 2/3 that of the effective chimney height
    c) effective chimney height
    d) 15 ft

2973. Cold air returns from bathrooms should be:
    a) same size as the hot air inlet
    b) 1" up from the floor
    c) omitted
    d) supplied with a damper

2974. Smoke pipe, class B vent must have a rise of:
    a) ¼" per inch
    b) ½" per inch
    c) ¼" per foot
    d) ½" per foot

2975. Smoke pipe, class B vent, the gauge would be:
   a) 26 ga
   b) 28 ga
   c) 30 ga
   d) none of the above
2976. Portable ladders can not be over:
   a) 15 ft
   b) 20 ft
   c) 25 ft
   d) 30 ft
2977. Portable ladders must have how many rungs projecting above the roof?
   a) 1
   b) 2
   c) 3
   d) 4
2978. The setting of the limit control on a gravity warm air furnace shall not exceed:
   a) 160
   b) 200
   c) 250
   d) 300
2979. According to BOCA regulations, to be an appliance of low heat, the products of combustion at the point of entrance to the flue must be:
   a) 500 to 1100°F
   b) 1000°F or less
   c) 1000 to 2000°F
   d) 2000 to 3000°F
2980. According to BOCA regulations, to be an appliance of medium heat, the products of combustion at the point of entrance to the flue must be:
   a) 500 to 1000°F
   b) 1000 to 2000°F
   c) 2000 to 3000°F
   d) 3000 to 4000°F
2981. According to BOCA regulations, to be an appliance of high heat, the products of combustion at the point of entrance to the flue must be:
   a) 100 to 500°F
   b) 501 to 1000°F
   c) 1001 to 2000°F
   d) 2001 and up
2982. The location of outside air exhaust and intake openings shall be located a minimum of _____ feet from lot lines or buildings on the same lot.
   a) 5
   b) 10
   c) 15
   d) 20

3000 Questions & Answers

2983. The supports of ducts shall be supported with approved hangers at intervals not exceeding _____ feet.
   a) 8
   b) 10
   c) 12
   d) 14

2984. Installations exhausting more than _____ cfm shall be provided with make-up air.
   a) 50
   b) 100
   c) 200
   d) 450

2985. Appliances located in public garages, service stations, repair garages, shall be installed a minimum of how many feet for the floor.
   a) 5
   b) 6
   c) 8
   d) 10

2986. A 125 volt AC grounding-type outlet shall be available for all appliances. The outlet shall be located on the same level, within _____ feet of the appliance.
   a) 25
   b) 50
   c) 75
   d) 100

2987. The quantity of gas to be provided at each outlet shall be determined by the manufacturer's:
   a) size of the pipe at the gas valve
   b) Btu output of the unit
   c) Btu input of the unit
   d) none of the above

2988. The approximate gas input for a free-standing domestic range is:
   a) 4,000 Btu
   b) 12,500 Btu
   c) 25,000 Btu
   d) 65,000 Btu

2989. The approximate gas input for a domestic water heater (30 to 40 gal) is:
   a) 15,000 Btu
   b) 20,000 Btu
   c) 45,000 Btu
   d) 85,000 Btu

2990. The approximate gas input for a domestic clothes dryer would be:
   a) 10,000 Btu
   b) 15,000 Btu

c) 35,000 Btu
d) 65,000 Btu

2991. The threading of gas pipe of ½" to 1" will have:
    a) 8 threads
    b) 9 threads
    c) 10 threads
    d) 11 threads

2992. The threading of gas pipe of 1¼" to 2" will have:
    a) 11 threads
    b) 12 threads
    c) 13 threads
    d) 14 threads

2993. The threading of gas pipe of 2½" to 3" will have:
    a) 11 threads
    b) 12 threads
    c) 13 threads
    d) 14 threads

2994. No gas pipe smaller than _____ inch iron pipe size shall be used in any concealed location.
    a) 1/4"
    b) 3/8"
    c) 1/2"
    d) 3/4"

2995. Gas piping outlets shall extend at least _____ through finished ceilings and walls.
    a) 1"
    b) 2"
    c) 3"
    d) 6"

2996. Gas piping outlets shall extend at least _____ through floors.
    a) 1"
    b) 2"
    c) 3"
    d) 6"

2997. When testing piping for tightness, use:
    a) oxygen
    b) oil
    c) air
    d) cool gas

2998. When bending steel gas pipe, you must:
    a) heat the pipe
    b) use a pipe bender
    c) both A & B
    d) none of the above

2999. You may install a gas pipe in an air duct if the gas pipe is:
   a) hard rubber gas line
   b) copper gas line
   c) steel gas line
   d) none of the above
3000. An oil furnace hanging from a 15 foot ceiling in a service garage requires the use of a:
   a) two-pipe, single-stage oil pump
   b) two-pipe, two-stage oil pump
   c) one-pipe, single-stage oil pump
   d) one-pipe, two-stage oil pump

SECTION ONE ANSWERS

1 A	51 B	101 C	151 C	201 B	251 A	301 A	351 B	401 C	451 A
2 C	52 C	102 A	152 B	202 C	252 B	302 C	352 A	402 C	452 A
3 B	53 A	103 B	153 C	203 C	253 B	303 C	353 C	403 C	453 D
4 B	54 B	104 C	154 A	204 C	254 B	304 B	354 B	404 A	454 A
5 C	55 A	105 C	155 B	205 C	255 B	305 A	355 A	405 C	455 A
6 A	56 A	106 C	156 B	206 B	256 C	306 B	356 B	406 C	456 B
7 B	57 A	107 C	157 B	207 A	257 C	307 C	357 B	407 B	457 D
8 B	58 B	108 C	158 A	208 A	258 C	308 A	358 B	408 C	458 A
9 B	59 A	109 B	159 B	209 A	259 B	309 B	359 A	409 C	459 B
10 C	60 A	110 C	160 C	210 B	260 A	310 B	360 B	410 C	460 A
11 C	61 A	111 C	161 B	211 B	261 A	311 C	361 B	411 B	461 D
12 C	62 A	112 B	162 B	212 A	262 B	312 A	362 C	412 B	462 D
13 B	63 A	113 C	163 C	213 A	263 B	313 A	363 C	413 B	463 C
14 A	64 A	114 C	164 A	214 A	264 C	314 B	364 B	414 C	464 A
15 A	65 B	115 B	165 B	215 A	265 B	315 B	365 A	415 B	465 B
16 C	66 A	116 B	166 B	216 B	266 C	316 A	366 B	416 A	466 C
17 A	67 B	117 A	167 C	217 C	267 B	317 C	367 C	417 A	467 A
18 A	68 C	118 A	168 A	218 B	268 B	318 C	368 C	418 A	468 B
19 B	69 C	119 B	169 C	219 A	269 C	319 C	369 B	419 C	469 A
20 C	70 A	120 B	170 C	220 A	270 C	320 B	370 B	420 C	470 B
21 A	71 B	121 A	171 C	221 B	271 C	321 B	371 B	421 C	471 C
22 C	72 C	122 C	172 C	222 A	272 C	322 B	372 C	422 B	472 A
23 B	73 A	123 C	173 B	223 A	273 E	323 B	373 A	423 D	473 C
24 A	74 B	124 C	174 A	224 A	274 C	324 C	374 A	424 C	474 C
25 A	75 C	125 C	175 A	225 B	275 C	325 C	375 B	425 D	475 D
26 A	76 C	126 C	176 B	226 A	276 A	326 A	376 A	426 A	476 D
27 C	77 A	127 C	177 C	227 C	277 C	327 A	377 B	427 B	477 D
28 B	78 B	128 B	178 A	228 A	278 A	328 A	378 C	428 A	478 A
29 A	79 B	129 C	179 B	229 A	279 B	329 C	379 C	429 B	479 A
30 B	80 B	130 A	180 A	230 B	280 B	330 C	380 C	430 B	480 D
31 B	81 B	131 A	181 C	231 B	281 C	331 B	381 B	431 A	481 D
32 C	82 A	132 B	182 C	232 B	282 C	332 C	382 C	432 B	482 C
33 B	83 B	133 C	183 A	233 C	283 B	333 A	383 C	433 A	483 B
34 B	84 C	134 B	184 C	234 C	284 C	334 C	384 B	434 B	484 A
35 A	85 B	135 C	185 B	235 A	285 C	335 A	385 A	435 D	485 D
36 C	86 A	136 B	186 A	236 B	286 A	336 A	386 C	436 B	486 A
37 A	87 B	137 B	187 A	237 C	287 B	337 B	387 A	437 B	487 B
38 C	88 B	138 B	188 A	238 A	288 A	338 A	388 B	438 A	488 B
39 B	89 B	139 A	189 A	239 B	289 A	339 C	389 C	439 D	489 C
40 C	90 A	140 B	190 B	240 A	290 B	340 A	390 A	440 A	490 C
41 B	91 B	141 C	191 A	241 A	291 A	341 A	391 B	441 C	491 A
42 B	92 C	142 B	192 A	242 C	292 A	342 C	392 B	442 C	492 C
43 B	93 C	143 C	193 B	243 C	293 C	343 C	393 C	443 B	493 C
44 C	94 A	144 A	194 B	244 A	294 A	344 C	394 A	444 A	494 C
45 C	95 C	145 B	195 A	245 A	295 B	345 C	395 A	445 C	495 B
46 A	96 A	146 B	196 B	246 B	296 C	346 B	396 A	446 D	496 C
47 C	97 B	147 B	197 A	247 C	297 C	347 C	397 C	447 B	497 C
48 C	98 C	148 B	198 A	248 B	298 A	348 B	398 B	448 D	498 D
49 A	99 C	149 C	199 B	249 B	299 A	349 A	399 C	449 A	499 B
50 A	100 A	150 A	200 B	250 C	300 B	350 C	400 B	450 C	500 C

Dundas

501 A	551 B	601 A	651 A	701 B	751 A	801 B	851 A	901 A	951 B
502 A	552 A	602 B	652 B	702 B	752 D	802 C	852 B	902 B	952 B
503 B	553 A	603 C	653 B	703 D	753 D	803 A	853 B	903 A	953 A
504 B	554 B	604 A	654 B	704 B	754 B	804 D	854 C	904 C	954 B
505 C	555 C	605 B	655 A	705 A	755 B	805 C	855 B	905 C	955 A
506 A	556 C	606 A	656 A	706 B	756 D	806 D	856 A	906 B	956 C
507 D	557 C	607 B	657 B	707 D	757 B	807 D	857 A	907 A	957 C
508 B	558 A	608 B	658 B	708 C	758 A	808 A	858 B	908 B	958 D
509 D	559 C	609 A	659 C	709 A	759 B	809 D	859 B	909 A	959 B
510 B	560 A	610 B	660 B	710 C	760 A	810 D	860 A	910 C	960 A
511 C	561 A	611 B	661 C	711 C	761 B	811 B	861 C	911 D	961 A
512 D	562 B	612 A	662 C	712 B	762 C	812 C	862 B	912 A	962 C
513 D	563 A	613 A	663 A	713 D	763 C	813 A	863 B	913 C	963 C
514 C	564 C	614 A	664 A	714 C	764 B	814 C	864 C	914 A	964 B
515 A	565 B	615 A	665 B	715 A	765 D	815 C	865 A	915 B	965 B
516 B	566 B	616 C	666 C	716 C	766 A	816 D	866 C	916 A	966 B
517 C	567 C	617 C	667 A	717 D	767 B	817 A	867 A	917 C	967 C
518 C	568 B	618 A	668 A	718 D	768 D	818 A	868 B	918 B	968 C
519 B	569 A	619 A	669 D	719 A	769 B	819 B	869 B	919 B	969 B
520 A	570 B	620 C	670 B	720 D	770 C	820 A	870 C	920 A	970 D
521 C	571 B	621 C	671 A	721 C	771 A	821 A	871 A	921 D	971 C
522 A	572 A	622 A	672 B	722 C	772 D	822 C	872 C	922 B	972 A
523 A	573 B	623 A	673 C	723 B	773 A	823 C	873 A	923 D	973 B
524 A	574 C	624 C	674 C	724 D	774 A	824 B	874 A	924 B	974 D
525 C	575 C	625 B	675 A	725 D	775 D	825 B	875 C	925 A	975 B
526 B	576 B	626 B	676 B	726 A	776 D	826 A	876 A	926 B	976 D
527 C	577 B	627 A	677 C	727 B	777 D	827 B	877 B	927 B	977 D
528 A	578 B	628 A	678 B	728 C	778 C	828 A	878 B	928 B	978 C
529 D	579 C	629 B	679 B	729 B	779 D	829 B	879 C	929 D	979 D
530 B	580 B	630 B	680 A	730 C	780 A	830 D	880 A	930 B	980 D
531 C	581 B	631 B	681 A	731 B	781 B	831 B	881 C	931 A	981 B
532 B	582 C	632 A	682 B	732 D	782 A	832 A	882 B	932 C	982 A
533 A	583 C	633 B	683 B	733 C	783 B	833 A	883 C	933 A	983 B
534 C	584 B	634 C	684 B	734 A	784 C	834 B	884 C	934 B	984 B
535 C	585 C	635 C	685 B	735 C	785 D	835 A	885 A	935 D	985 D
536 A	586 B	636 C	686 B	736 C	786 D	836 B	886 C	936 C	986 D
537 A	587 A	637 C	687 B	737 B	787 B	837 B	887 B	937 D	987 C
538 B	588 B	638 C	688 B	738 A	788 A	838 B	888 A	938 A	988 C
539 C	589 B	639 B	689 B	739 C	789 C	839 A	889 A	939 A	989 C
540 A	590 A	640 C	690 A	740 D	790 B	840 B	890 A	940 A	990 C
541 A	591 A	641 C	691 A	741 A	791 D	841 B	891 C	941 B	991 C
542 A	592 A	642 A	692 C	742 D	792 D	842 A	892 B	942 C	992 D
543 B	593 A	643 C	693 A	743 C	793 B	843 A	893 A	943 A	993 C
544 C	594 B	644 B	694 B	744 A	794 A	844 A	894 A	944 A	994 C
545 C	595 C	645 B	695 D	745 C	795 C	845 A	895 C	945 B	995 C
546 A	596 C	646 B	696 C	746 B	796 B	846 A	896 A	946 B	996 C
547 A	597 C	647 C	697 C	747 A	797 A	847 A	897 A	947 A	997 D
548 B	598 B	648 A	698 D	748 D	798 C	848 B	898 C	948 A	998 B
549 A	599 C	649 A	699 A	749 C	799 D	849 B	899 A	949 D	999 C
550 C	600 A	650 C	700 B	750 A	800 C	850 C	900 C	950 A	1000 D

3000 Questions & Answers

1001 A	1051 A	1101 A	1151 B	1201 C	1251 C	1301 B	1351 A	1401 A	1451 A
1002 D	1052 A	1102 B	1152 C	1202 B	1252 A	1302 A	1352 B	1402 C	1452 C
1003 B	1053 A	1103 A	1153 A	1203 A	1253 A	1303 A	1353 C	1403 A	1453 B
1004 A	1054 A	1104 A	1154 A	1204 A	1254 B	1304 C	1354 B	1404 A	1454 C
1005 C	1055 A	1105 A	1155 A	1205 B	1255 B	1305 A	1355 C	1405 D	1455 A
1006 D	1056 B	1106 A	1156 B	1206 A	1256 C	1306 A	1356 B	1406 B	1456 A
1007 C	1057 B	1107 B	1157 C	1207 A	1257 C	1307 A	1357 D	1407 C	1457 A
1008 B	1058 B	1108 A	1158 A	1208 A	1258 A	1308 A	1358 A	1408 A	1458 B
1009 B	1059 B	1109 B	1159 C	1209 B	1259 B	1309 B	1359 D	1409 A	1459 A
1010 D	1060 B	1110 B	1160 A	1210 C	1260 C	1310 D	1360 B	1410 C	1460 B
1011 A	1061 B	1111 B	1161 B	1211 C	1261 C	1311 A	1361 D	1411 C	1461 D
1012 C	1062 B	1112 A	1162 C	1212 A	1262 A	1312 D	1362 A	1412 A	1462 D
1013 A	1063 A	1113 A	1163 C	1213 B	1263 B	1313 D	1363 B	1413 E	1463 C
1014 B	1064 A	1114 B	1164 C	1214 B	1264 D	1314 A	1364 A	1414 E	1464 B
1015 A	1065 B	1115 B	1165 B	1215 C	1265 D	1315 B	1365 B	1415 C	1465 D
1016 B	1066 B	1116 A	1166 C	1216 C	1266 C	1316 D	1366 B	1416 E	1466 A
1017 A	1067 A	1117 B	1167 C	1217 B	1267 B	1317 C	1367 A	1417 B	1467 A
1018 B	1068 B	1118 B	1168 B	1218 B	1268 B	1318 B	1368 D	1418 E	1468 C
1019 A	1069 A	1119 B	1169 A	1219 C	1269 B	1319 A	1369 B	1419 D	1469 C
1020 C	1070 A	1120 A	1170 D	1220 A	1270 B	1320 B	1370 B	1420 D	1470 B
1021 C	1071 B	1121 D	1171 A	1221 A	1271 C	1321 B	1371 B	1421 B	1471 A
1022 A	1072 A	1122 A	1172 C	1222 C	1272 B	1322 C	1372 A	1422 D	1472 D
1023 B	1073 A	1123 B	1173 C	1223 B	1273 A	1323 D	1373 B	1423 D	1473 C
1024 C	1074 A	1124 B	1174 B	1224 D	1274 C	1324 B	1374 A	1424 C	1474 A
1025 A	1075 B	1125 A	1175 C	1225 B	1275 D	1325 A	1375 B	1425 C	1475 C
1026 A	1076 B	1126 B	1176 A	1226 B	1276 D	1326 C	1376 B	1426 A	1476 A
1027 B	1077 A	1127 B	1177 A	1227 B	1277 C	1327 A	1377 D	1427 C	1477 C
1028 B	1078 A	1128 B	1178 B	1228 A	1278 A	1328 C	1378 B	1428 A	1478 B
1029 A	1079 A	1129 A	1179 B	1229 C	1279 A	1329 D	1379 D	1429 D	1479 C
1030 A	1080 B	1130 B	1180 B	1230 B	1280 A	1330 B	1380 C	1430 E	1480 B
1031 B	1081 B	1131 A	1181 B	1231 B	1281 B	1331 A	1381 A	1431 B	1481 A
1032 A	1082 B	1132 B	1182 B	1232 C	1282 B	1332 C	1382 A	1432 C	1482 A
1033 A	1083 B	1133 B	1183 C	1233 A	1283 B	1333 D	1383 A	1433 A	1483 B
1034 B	1084 B	1134 A	1184 C	1234 B	1284 B	1334 B	1384 C	1434 D	1484 C
1035 A	1085 A	1135 B	1185 C	1235 B	1285 D	1335 A	1385 B	1435 A	1485 D
1036 A	1086 B	1136 C	1186 B	1236 B	1286 A	1336 B	1386 A	1436 A	1486 C
1037 B	1087 B	1137 B	1187 C	1237 D	1287 A	1337 D	1387 C	1437 B	1487 A
1038 B	1088 A	1138 A	1188 C	1238 C	1288 B	1338 B	1388 C	1438 A	1488 A
1039 A	1089 A	1139 B	1189 B	1239 B	1289 D	1339 A	1389 B	1439 B	1489 A
1040 A	1090 B	1140 A	1190 A	1240 A	1290 A	1340 A	1390 B	1440 D	1490 B
1041 B	1091 A	1141 A	1191 A	1241 C	1291 C	1341 A	1391 B	1441 B	1491 C
1042 A	1092 A	1142 B	1192 B	1242 D	1292 A	1342 A	1392 D	1442 E	1492 E
1043 A	1093 A	1143 A	1193 B	1243 C	1293 D	1343 D	1393 D	1443 D	1493 E
1044 A	1094 A	1144 B	1194 C	1244 C	1294 C	1344 D	1394 C	1444 E	1494 E
1045 B	1095 A	1145 C	1195 C	1245 A	1295 B	1345 A	1395 B	1445 A	1495 C
1046 A	1096 B	1146 B	1196 B	1246 B	1296 A	1346 B	1396 C	1446 E	1496 A
1047 A	1097 B	1147 A	1197 B	1247 C	1297 B	1347 D	1397 C	1447 C	1497 C
1048 A	1098 B	1148 B	1198 C	1248 D	1298 C	1348 A	1398 A	1448 D	1498 A
1049 A	1099 B	1149 C	1199 C	1249 B	1299 A	1349 C	1399 B	1449 D	1499 C
1050 B	1100 B	1150 C	1200 A	1250 C	1300 B	1350 D	1400 D	1450 E	1500 C

SECTION TWO
ANSWERS

1501 B	1526 B	1551 C	1576 C		1701 B	1726 A	1751 A	1776 A
1502 C	1527 D	1552 D	1577 C		1702 C	1727 B	1752 A	1777 B
1503 C	1528 B	1553 C	1578 D		1703 B	1728 B	1753 A	1778 C
1504 B	1529 C	1554 B	1579 C		1704 C	1729 A	1754 A	1779 D
1505 D	1530 C	1555 A	1580 A		1705 B	1730 A	1755 A	1780 A
1506 A	1531 A	1556 B	1581 E		1706 A	1731 C	1756 A	1781 C
1507 D	1532 D	1557 C	1582 A		1707 A	1732 A	1757 A	1782 A
1508 A	1533 D	1558 B	1583 C		1708 B	1733 C	1758 B	1783 A
1509 B	1534 A	1559 C	1584 E		1709 B	1734 A	1759 A	1784 B
1510 C	1535 B	1560 C	1585 A		1710 C	1735 B	1760 A	1785 B
1511 D	1536 C	1561 B	1586 E		1711 B	1736 A	1761 C	1786 B
1512 C	1537 C	1562 B	1587 B		1712 C	1737 B	1762 A	1787 C
1513 D	1538 C	1563 B	1588 E		1713 A	1738 A	1763 C	1788 C
1514 C	1539 B	1564 D	1589 A		1714 C	1739 B	1764 A	1789 B
1515 C	1540 A	1565 B	1590 A		1715 C	1740 A	1765 A	1790 A
1516 A	1541 A	1566 D	1591 C		1716 B	1741 B	1766 A	1791 B
1517 B	1542 D	1567 D	1592 C		1717 B	1742 A	1767 C	1792 B
1518 C	1543 C	1568 C	1593 C		1718 A	1743 A	1768 C	1793 C
1519 C	1544 B	1569 A	1594 A		1719 C	1744 C	1769 B	1794 A
1520 C	1545 C	1570 D	1595 B		1720 C	1745 B	1770 B	1795 B
1521 D	1546 A	1571 E	1596 B		1721 B	1746 C	1771 A	1796 A
1522 C	1547 B	1572 C	1597 A		1722 B	1747 A	1772 A	1797 B
1523 B	1548 B	1573 B	1598 D		1723 C	1748 B	1773 B	1798 B
1524 C	1549 C	1574 B	1599 A		1724 A	1749 A	1774 B	1799 B
1525 B	1550 C	1575 C	1600 C		1725 C	1750 A	1775 C	1800 A
1601 D	1626 C	1651 B	1676 D		1801 A	1826 B	1851 A	1876 A
1602 C	1627 D	1652 C	1677 A		1802 B	1827 C	1852 B	1877 B
1603 B	1628 C	1653 B	1678 C		1803 B	1828 C	1853 A	1878 B
1604 D	1629 A	1654 D	1679 B		1804 A	1829 A	1854 C	1879 C
1605 B	1630 C	1655 A	1680 D		1805 C	1830 B	1855 B	1880 A
1606 D	1631 A	1656 D	1681 B		1806 B	1831 A	1856 B	1881 C
1607 E	1632 B	1657 D	1682 D		1807 B	1832 A	1857 C	1882 A
1608 D	1633 D	1658 B	1683 A		1808 B	1833 C	1858 A	1883 C
1609 E	1634 D	1659 B	1684 D		1809 B	1834 B	1859 A	1884 B
1610 E	1635 C	1660 B	1685 A		1810 B	1835 A	1860 A	1885 B
1611 D	1636 A	1661 B	1686 B		1811 B	1836 B	1861 B	1886 A
1612 B	1637 B	1662 C	1687 B		1812 A	1837 C	1862 A	1887 A
1613 C	1638 C	1663 C	1688 D		1813 A	1838 A	1863 B	1888 B
1614 E	1639 E	1664 B	1689 B		1814 B	1839 B	1864 A	1889 A
1615 E	1640 B	1665 C	1690 B		1815 B	1840 A	1865 A	1890 A
1616 B	1641 C	1666 C	1691 C		1816 B	1841 A	1866 C	1891 A
1617 A	1642 D	1667 A	1692 B		1817 B	1842 C	1867 A	1892 B
1618 C	1643 B	1668 D	1693 B		1818 C	1843 A	1868 B	1893 A
1619 D	1644 C	1669 A	1694 A		1819 B	1844 B	1869 C	1894 B
1620 C	1645 C	1670 B	1695 D		1820 A	1845 A	1870 C	1895 A
1621 D	1646 A	1671 B	1696 C		1821 B	1846 B	1871 B	1896 A
1622 A	1647 C	1672 C	1697 C		1822 B	1847 B	1872 B	1897 B
1623 B	1648 B	1673 B	1698 C		1823 B	1848 B	1873 C	1898 A
1624 C	1649 A	1674 A	1699 B		1824 A	1849 A	1874 B	1899 B
1625 C	1650 C	1675 C	1700 C		1825 A	1850 B	1875 B	1900 A

Dundas

1901 C	1926 B	1951 C	1976 A	2101 A	2126 B	2151 B	2176 C
1902 C	1927 C	1952 B	1977 A	2102 A	2127 C	2152 A	2177 C
1903 C	1928 B	1953 A	1978 A	2103 C	2128 A	2153 B	2178 B
1904 C	1929 B	1954 A	1979 B	2104 B	2129 B	2154 A	2179 B
1905 C	1930 A	1955 C	1980 B	2105 A	2130 B	2155 A	2180 B
1906 A	1931 C	1956 A	1981 B	2106 B	2131 A	2156 B	2181 B
1907 B	1932 A	1957 A	1982 B	2107 A	2132 A	2157 A	2182 B
1908 A	1933 A	1958 C	1983 A	2108 C	2133 A	2158 C	2183 B
1909 C	1934 C	1959 A	1984 A	2109 C	2134 C	2159 B	2184 B
1910 C	1935 A	1960 B	1985 A	2110 B	2135 C	2160 B	2185 B
1911 A	1936 C	1961 B	1986 C	2111 C	2136 C	2161 B	2186 A
1912 B	1937 B	1962 C	1987 B	2112 A	2137 B	2162 A	2187 A
1913 B	1938 B	1963 C	1988 D	2113 B	2138 A	2163 B	2188 B
1914 B	1939 B	1964 C	1989 B	2114 C	2139 C	2164 B	2189 B
1915 C	1940 B	1965 C	1990 C	2115 B	2140 B	2165 A	2190 C
1916 C	1941 B	1966 A	1991 A	2116 C	2141 B	2166 A	2191 A
1917 C	1942 A	1967 B	1992 B	2117 A	2142 A	2167 C	2192 A
1918 B	1943 A	1968 A	1993 A	2118 C	2143 C	2168 A	2193 C
1919 A	1944 B	1969 C	1994 B	2119 B	2144 B	2169 C	2194 A
1920 A	1945 A	1970 C	1995 C	2120 B	2145 C	2170 B	2195 A
1921 B	1946 B	1971 C	1996 C	2121 A	2146 B	2171 C	2196 A
1922 B	1947 A	1972 B	1997 A	2122 A	2147 C	2172 C	2197 B
1923 B	1948 B	1973 C	1998 A	2123 C	2148 B	2173 D	2198 A
1924 C	1949 B	1974 A	1999 A	2124 A	2149 C	2174 A	2199 B
1925 A	1950 C	1975 B	2000 C	2125 C	2150 C	2175 C	2200 B
2001 A	2026 A	2051 A	2076 A	2201 B	2226 C	2251 A	2276 B
2002 A	2027 A	2052 B	2077 A	2202 A	2227 B	2252 B	2277 C
2003 C	2028 C	2053 A	2078 C	2203 A	2228 A	2253 B	2278 C
2004 C	2029 C	2054 C	2079 B	2204 C	2229 B	2254 B	2279 C
2005 A	2030 B	2055 B	2080 A	2205 A	2230 B	2255 A	2280 A
2006 C	2031 B	2056 A	2081 A	2206 B	2231 A	2256 A	2281 C
2007 C	2032 A	2057 A	2082 B	2207 C	2232 A	2257 B	2282 A
2008 A	2033 A	2058 C	2083 B	2208 A	2233 C	2258 C	2283 B
2009 A	2034 B	2059 B	2084 B	2209 C	2234 A	2259 A	2284 A
2010 B	2035 C	2060 C	2085 B	2210 A	2235 C	2260 B	2285 A
2011 A	2036 B	2061 C	2086 A	2211 C	2236 B	2261 B	2286 A
2012 C	2037 B	2062 B	2087 A	2212 A	2237 A	2262 A	2287 C
2013 A	2038 B	2063 C	2088 A	2213 A	2238 D	2263 B	2288 C
2014 A	2039 C	2064 B	2089 B	2214 A	2239 C	2264 C	2289 B
2015 B	2040 A	2065 B	2090 B	2215 B	2240 B	2265 A	2290 B
2016 B	2041 C	2066 C	2091 C	2216 A	2241 A	2266 A	2291 A
2017 C	2042 A	2067 B	2092 C	2217 A	2242 C	2267 B	2292 B
2018 A	2043 A	2068 A	2093 C	2218 A	2243 C	2268 B	2293 B
2019 A	2044 B	2069 A	2094 C	2219 B	2244 A	2269 C	2294 A
2020 A	2045 A	2070 C	2095 A	2220 B	2245 B	2270 B	2295 B
2021 A	2046 C	2071 B	2096 C	2221 C	2246 B	2271 B	2296 C
2022 A	2047 A	2072 B	2097 A	2222 C	2247 B	2272 B	2297 B
2023 A	2048 B	2073 C	2098 A	2223 B	2248 A	2273 A	2298 C
2024 B	2049 B	2074 A	2099 A	2224 A	2249 B	2274 C	2299 B
2025 B	2050 C	2075 C	2100 C	2225 A	2250 A	2275 B	2300 C

3000 Questions & Answers

2301 A	2326 A	2351 A	2376 A	2501 B	2526 A	2551 A	2576 B
2302 B	2327 B	2352 A	2377 B	2502 B	2527 C	2552 A	2577 C
2303 B	2328 B	2353 A	2378 A	2503 B	2528 B	2553 A	2578 B
2304 B	2329 C	2354 A	2379 C	2504 B	2529 A	2554 C	2579 B
2305 B	2330 C	2355 A	2380 C	2505 B	2530 A	2555 B	2580 B
2306 A	2331 B	2356 A	2381 A	2506 A	2531 C	2556 A	2581 B
2307 A	2332 A	2357 B	2382 C	2507 C	2532 A	2557 B	2582 A
2308 B	2333 A	2358 B	2383 C	2508 A	2533 A	2558 B	2583 C
2309 A	2334 C	2359 A	2384 C	2509 B	2534 B	2559 A	2584 B
2310 B	2335 A	2360 A	2385 C	2510 B	2535 A	2560 A	2585 B
2311 B	2336 A	2361 B	2386 B	2511 B	2536 B	2561 A	2586 C
2312 B	2337 B	2362 A	2387 C	2512 C	2537 A	2562 A	2587 B
2313 D	2338 C	2363 B	2388 A	2513 A	2538 C	2563 C	2588 A
2314 C	2339 B	2364 C	2389 C	2514 A	2539 B	2564 B	2589 B
2315 B	2340 A	2365 A	2390 A	2515 A	2540 C	2565 B	2590 A
2316 B	2341 C	2366 A	2391 A	2516 A	2541 B	2566 C	2591 B
2317 A	2342 C	2367 A	2392 A	2517 A	2542 C	2567 B	2592 A
2318 B	2343 A	2368 C	2393 B	2518 B	2543 B	2568 B	2593 B
2319 A	2344 A	2369 C	2394 B	2519 B	2544 C	2569 A	2594 A
2320 A	2345 B	2370 C	2395 A	2520 B	2545 B	2570 B	2595 B
2321 A	2346 A	2371 A	2396 A	2521 B	2546 B	2571 A	2596 A
2322 A	2347 B	2372 C	2397 C	2522 B	2547 C	2572 B	2597 B
2323 B	2348 A	2373 C	2398 A	2523 A	2548 A	2573 B	2598 C
2324 C	2349 A	2374 B	2399 A	2524 C	2549 B	2574 C	2599 B
2325 C	2350 A	2375 A	2400 A	2525 C	2550 C	2575 C	2600 A
2401 A	2426 C	2451 B	2476 B	2601 C	2626 A	2651 C	2676 A
2402 C	2427 B	2452 A	2477 A	2602 C	2627 A	2652 A	2677 A
2403 A	2428 B	2453 B	2478 B	2603 B	2628 B	2653 C	2678 A
2404 A	2429 A	2454 C	2479 A	2604 D	2629 D	2654 B	2679 B
2405 C	2430 A	2455 B	2480 B	2605 B	2630 C	2655 C	2680 B
2406 A	2431 B	2456 A	2481 A	2606 A	2631 D	2656 A	2681 B
2407 B	2432 B	2457 B	2482 A	2607 A	2632 B	2657 B	2682 B
2408 A	2433 B	2458 C	2483 B	2608 D	2633 A	2658 B	2683 A
2409 A	2434 C	2459 B	2484 A	2609 B	2634 A	2659 C	2684 B
2410 A	2435 C	2460 B	2485 A	2610 B	2635 A	2660 D	2685 C
2411 C	2436 A	2461 A	2486 C	2611 A	2636 A	2661 A	2686 A
2412 A	2437 C	2462 C	2487 B	2612 A	2637 A	2662 B	2687 D
2413 B	2438 C	2463 B	2488 A	2613 B	2638 C	2663 B	2688 B
2414 B	2439 C	2464 B	2489 B	2614 C	2639 A	2664 A	2689 A
2415 B	2440 A	2465 A	2490 A	2615 D	2640 C	2665 C	2690 B
2416 A	2441 C	2466 A	2491 B	2616 C	2641 D	2666 A	2691 C
2417 A	2442 A	2467 C	2492 B	2617 A	2642 C	2667 A	2692 C
2418 B	2443 A	2468 B	2493 A	2618 D	2643 B	2668 D	2693 A
2419 C	2444 C	2469 C	2494 A	2619 C	2644 C	2669 C	2694 C
2420 B	2445 A	2470 B	2495 B	2620 D	2645 B	2670 A	2695 C
2421 C	2446 A	2471 B	2496 A	2621 D	2646 C	2671 B	2696 C
2422 B	2447 B	2472 A	2497 B	2622 A	2647 B	2672 A	2697 C
2423 A	2448 A	2473 A	2498 B	2623 A	2648 A	2673 C	2698 C
2424 A	2449 B	2474 A	2499 B	2624 A	2649 D	2674 B	2699 C
2425 A	2450 A	2475 A	2500 A	2625 B	2650 A	2675 B	2700 D

Dundas

2701 A	2726 C	2751 B	2776 B		2901 D	2926 B	2951 D	2976 B
2702 C	2727 D	2752 D	2777 B		2902 B	2927 B	2952 C	2977 C
2703 B	2728 B	2753 B	2778 B		2903 C	2928 D	2953 B	2978 D
2704 C	2729 A	2754 B	2779 C		2904 D	2929 C	2954 B	2979 B
2705 D	2730 A	2755 A	2780 D		2905 C	2930 A	2955 C	2980 B
2706 B	2731 C	2756 B	2781 B		2906 C	2931 C	2956 A	2981 D
2707 C	2732 A	2757 C	2782 B		2907 D	2932 D	2957 C	2982 B
2708 A	2733 D	2758 A	2783 C		2908 B	2933 B	2958 C	2983 B
2709 B	2734 A	2759 A	2784 B		2909 C	2934 B	2959 D	2984 C
2710 B	2735 C	2760 B	2785 A		2910 B	2935 B	2960 C	2985 C
2711 A	2736 B	2761 C	2786 B		2911 A	2936 D	2961 D	2986 D
2712 A	2737 A	2762 B	2787 A		2912 B	2937 D	2962 A	2987 C
2713 B	2738 D	2763 A	2788 B		2913 D	2938 C	2963 A	2988 D
2714 D	2739 D	2764 B	2789 D		2914 B	2939 C	2964 B	2989 C
2715 A	2740 A	2765 B	2790 B		2915 C	2940 B	2965 A	2990 C
2716 D	2741 A	2766 A	2791 B		2916 D	2941 C	2966 B	2991 C
2717 B	2742 C	2767 A	2792 A		2917 B	2942 C	2967 D	2992 A
2718 C	2743 B	2768 D	2793 B		2918 B	2943 D	2968 C	2993 B
2719 C	2744 A	2769 C	2794 D		2919 A	2944 B	2969 A	2994 C
2720 A	2745 D	2770 B	2795 C		2920 C	2945 D	2970 B	2995 A
2721 B	2746 C	2771 D	2796 B		2921 C	2946 B	2971 A	2996 B
2722 C	2747 A	2772 C	2797 B		2922 C	2947 B	2972 B	2997 C
2723 A	2748 B	2773 B	2798 A		2923 C	2948 C	2973 C	2998 D
2724 D	2749 D	2774 A	2799 B		2924 B	2949 B	2974 C	2999 D
2725 B	2750 C	2775 C	2800 A		2925 A	2950 A	2975 A	3000 B
2801 A	2826 C	2851 B	2876 A					
2802 C	2827 A	2852 A	2877 D					
2803 A	2828 C	2853 D	2878 A					
2804 B	2829 B	2854 A	2879 B					
2805 B	2830 C	2855 C	2880 A					
2806 B	2831 C	2856 D	2881 C					
2807 B	2832 A	2857 B	2882 C					
2808 C	2833 B	2858 C	2883 C					
2809 A	2834 A	2859 A	2884 B					
2810 B	2835 C	2860 A	2885 C					
2811 B	2836 C	2861 A	2886 A					
2812 B	2837 B	2862 A	2887 A					
2813 C	2838 A	2863 B	2888 A					
2814 A	2839 A	2864 B	2889 B					
2815 A	2840 A	2865 C	2890 D					
2816 B	2841 B	2866 C	2891 C					
2817 A	2842 B	2867 A	2892 B					
2818 B	2843 A	2868 C	2893 A					
2819 B	2844 B	2869 A	2894 C					
2820 C	2845 A	2870 C	2895 B					
2821 C	2846 A	2871 A	2896 D					
2822 C	2847 A	2872 B	2897 B					
2823 C	2848 C	2873 B	2898 A					
2824 B	2849 C	2874 D	2899 D					
2825 A	2850 C	2875 A	2900 D					

Open New Doors With ...

HOW TO CLOSE
A Guide to Residential Heating & Cooling Sales

Times are changing and so are selling techniques. To be a successful salesman today takes professionalism, skill and work. Most importantly, your customers expect you to be knowledgeable about your product and how it can help them. This book prepares you for every possible situation you're likely to encounter. Anyone can become a super salesman by following the practical guidelines presented in this book. Author, Clayton Carrico who founded and operated a successful heating/cooling service company for twenty-seven years, will help you:

- identify the 7 customer types and deal with each of them
- overcome customer objections, especially price
- read between the lines of customer objections
- convince the customer of your product's benefits

Also included for your own use are survey sheets to fill out on the job site. These will help you record all the essential data necessary to properly size and evaluate heating/cooling equipment and they will also increase your customer's confidence in you as a professional.

Be Prepared to Answer Any Question Your Customer Asks and
DON'T LOSE ANOTHER SALE!

The solution to ice maker servicing is crystal clear...

THE HVAC/R PROFESSIONAL'S FIELD GUIDE TO
ICE MACHINE SERVICE

Diversify your skills and expand your reputation as a service technician. This manufacturer specific book gives you a solid understanding of ice machines so you can become a skilled ice maker repairman.

8½ x 11 • 748 pages

This practical guide takes the mystery out of commercial ice makers. The roots of ice maker service, maintenance and problem solving (as well as others) are presented in this guide to help you unlock your potential as an ice maker service specialist. Earn more money by exploring a whole new outlet for your skills and gain a more prestigious reputation as a knowledgeable and helpful service technician.

Many customers depend on their servicemen to educate and advise them on ice makers. Help your customers and yourself by learning the fundamentals of commercial ice machines with *Ice Machine Service* by Richard Jazwin

---

Manufacturer specific service manuals for select light commercial models (200-800 lbs ice/day) provide the specific troubleshooting and installation guidelines today's hvac/r service professional needs to excel in this competitive field. Includes information from:

Crystal Tips • Ice-O-Matic • Kold-Draft
Manitowoc • Remcor • Ross-Temp • Scotsman

Technical Book Division • Business News Publishing Company
P. O. Box 2600 • Troy • Michigan • 48007

NOW AVAILABLE FROM BNP

TROUBLESHOOTING & SERVICING AIR CONDITIONING EQUIPMENT

by
Don Swenson

This practical, concise book presents an in-depth study of troubleshooting and repairing **air conditioning and heat pump equipment**. The material is organized to progress from general principles of safety, tools, chart interpretation, instrumentation and troubleshooting, to descriptions of specific systems and components. The rest of the text is then devoted to applying this general information to commonly occurring problems in each of the component parts and subsystems of air conditioning, heat pump and refrigeration units.

Special features of this text include:

- **A focused approach:** The book clearly and concisely covers essentials and procedures.

- **Concept review:** Early chapters provide a review of system operation and provide basic but essential information for diagnosing and servicing.

- **Tool and safety information:** Excellent coverage is provided concerning the hazards and safety procedures required when working with air conditioning and heat pump equipment.

- **Troubleshooting charts:** Extensive troubleshooting and servicing flow-charts provide concise summaries of procedures described in the text and serve as quick diagnosing reference for job-site situations.

- **Systematic coverage:** A general systematic method of troubleshooting air conditioning and heat pump equipment is presented.

- **Review problems:** A number of problems and questions are included at the end of each chapter to aid in understanding the material presented when used in a training environment.

Technical Book Division • Business News Publishing Company
P. O. Box 2600 • Troy • Michigan • 48007

THE BEST REFERENCE AVAILABLE

THE HVAC/R PROFESSIONAL'S FIELD GUIDE TO
HEAT PUMP SYSTEMS & SERVICE

Finally, all the information you need to install, repair and maintain heat pumps has been compiled into one up-to-date, comprehensive guide. *Heat Pump Systems & Service* is the first in a new series of professional service guides from the BNP Technical Book Division — leading hvac/r publisher.

8½ x 11 • 380 pages

The text contains 8 comprehensive sections describing heat pump origination, operation, and components, all supplemented with illustrations, charts, troubleshooting tips, and repair strategies. The appendix contains manufacturers' data for selecte, top selling models. Includes instructions for—

- **unit placement**
- **installation & charging**
- **service & maintenance**
- **electrical connections & terminal ID**
- **accessories & replacement parts**

Knowing how specific units are configured makes service easier and more accurate, and improves your reputation as a trustworthy serviceman.

Never before has so much valuable, manufacturer specific heat pump information been collected in one handy source. Every service professional or tech-in-training will find this book absolutely indispensable.

Includes select manufacturer's data covering the top selling models from:

Carrier • Coleman • Comfortmaker • Day & Night, Payne
Goettl • Heil-Quaker • Lennox
Rheem • Sun Dial • Weathertron • York

Technical Book Division • Business News Publishing Company
P. O. Box 2600 • Troy • Michigan • 48007